Cooperative Management

The Book Series on Cooperative Management provides an invaluable forum for creative and scholarship work on cooperative economics, organizational, financial and marketing aspects of business cooperatives and development of cooperative communities throughout the Mediterranean region and worldwide. The main objectives of this book series are to advance knowledge related to cooperative entrepreneurship as well as to generate theoretical knowledge aiming to promoting research within various sectors wherein cooperatives operate (agriculture, banking, real estate, insurance, and other forms). Scholarly edited volumes and monographs should relate to one of these areas, should have a theoretical and/or empirical problem orientation, and should demonstrate innovation in theoretical and empirical analyses, methodologies, and applications. Analyses of cooperative economic problems and phenomena pertinent to managerial research, extension, and teaching (e.g., case studies) regarding cooperative entrepreneurship are equally encouraged.

Konstadinos Mattas • George Baourakis
Constantin Zopounidis • Christos Staboulis
Editors

Value Chain Dynamics in a Biodiverse Environment

Advances in Biodiversity, Sustainability, and Agri-food Supply Chain Development

Springer

Editors
Konstadinos Mattas
Department of Agricultural Economics
Aristotle University Of Thessaloniki
Thessaloniki, Greece

George Baourakis
Mediterranean Agronomic Institute
of Chania
Chania, Greece

Constantin Zopounidis
School of Production Engineering
and Management
Technical University of Crete
Chania, Greece

Christos Staboulis
Department of Agricultural Economics
Aristotle University of Thessaloniki
Thessaloniki, Greece

ISSN 2364-401X ISSN 2364-4028 (electronic)
Cooperative Management
ISBN 978-3-031-49844-2 ISBN 978-3-031-49845-9 (eBook)
https://doi.org/10.1007/978-3-031-49845-9

This Springer imprint is published by the registered company Springer Nature Switzerland AG
The registered company address is: Gewerbestrasse 11, 6330 Cham, Switzerland

Paper in this product is recyclable

Preface

The main objective of this volume is to document and advance knowledge related to practices and research initiatives that cover most of the current biodiversity, sustainability, and agri-food supply chain development. This will be of assistance in policy-making decisions in agriculture and the agri-food industry of the global economy and the subsequent prevailing societal and environmental conditions, as these are characterized by desertification and floods, disease outbreaks, increasing trade costs, and food insecurity.

It is commonly accepted that the agri-food supply chains are often characterized by poor sustainability performance. The various and heterogeneous stakeholders involved in the agri-food supply chain, the contradictory objectives and business mentalities of the actors, and the complicated processes and flows of materials/goods/information that take place throughout it are among the main factors for the low sustainability performance level. From this perspective, the first chapter involved in this book provides an end-to-end mapping of the Cretan vegetable supply chain, utilizing stakeholders' participatory approaches. The analysis highlights certain blockages of information flows and conflicting interactions among stakeholders, at the same time offering valuable insights and recommendations for future research and targeted policy-driven solutions.

Beyond the policy measures that foster the adoption of sustainable agri-food value chains, the trade of agri-food products remains a significant priority. Trade initiatives play a crucial role in promoting economic growth and development worldwide by reducing trade costs, improving access to global markets, and increasing the export of agricultural products. In line with this, the second chapter of this book provides an overview of trade competitiveness, revealing comparative advantages in the agri-food sector while focusing on the Latin America and Caribbean regions.

Cooperatives are member-oriented organizations, and their sustainable performance is highly influenced by the level of commitment of their members. Commitment is driven by member preferences for intra-organizational and strategic attributes of a cooperative. To this end, the work presented in the third chapter attempts to study empirically the relative importance that members attach to selected cooperative attributes. Knowledge of the existence of member segments and an understanding of their preferences may be useful to cooperative policy makers and managers to better evaluate the extent of member commitment and make more informed decisions of how a cooperative can service its membership needs.

Understanding factors that drive sustainable food consumption patterns is a crucial issue for the future well-being of humans, food systems sustainability, and environmental protection. Reinforcing sustainable diets based on environmentally friendly foods, changing food consumption habits, and increasing demand for organic food can reduce the food-related carbon footprint, mitigate the negative impacts of climate change, improve the quality of human life, and promote changes in the retail, distribution, and marketing functions of business. The fourth chapter of this book provides a critical review of the trend toward healthy and environmentally friendly food. The findings suggest that the established attitudes toward climate change, the preferences for healthier food products, and the social awareness toward environmentally friendly products seem to be the key factors to change purchase and consumption behavior.

The assessment of environmental degradation and its impact on socioeconomic conditions has gained significant importance for the long-term management of water resources and ecosystem sustainability. In this framework, the work presented in the fifth chapter of the book attempts to evaluate the sustainability of water utilization in selected areas of four African countries through the use of the Driving Forces-Pressure-Status-Impact-Response (DPSIR) model in a geographic information system (GIS) environment, aiming at contributing to the optimization of water resources management.

Value chain (VC)-based approaches have been appropriate for development in agricultural and food systems. The sixth chapter of this book indicates that approaching value chain selection from economic, social, and environmental perspectives thereby facilitates a successful start to the VC development cycle. Through a detailed review process, the chapter attempts to provide the guidelines for sustainable food systems which can ensure economic and social prosperity while preserving the environment and biodiversity for agri-food VC.

The seventh chapter examines and analyzes the financial situation of the brewery industry. To ensure the sustainability of the industry and understand the economic situation of brewery enterprises, financial analysis is crucial. By analyzing financial ratios, a company can gain insight into factors affecting economic utility, such as increasing profitability, reducing risk, and enhancing liquidity. As a result, it will be able to create a competitive advantage and follow a successful strategy.

The last chapter analyzes the value chain of carob flour production in a semi-mountainous region of Crete with significant economic, social, and cultural importance. The research identifies the main links of the value chain through interviews, regional government statistics, archival sources, and workshop results. Vulnerability analysis identified the factors of demographics, antagonizing cultivations, and international competition as the main reasons for the reduced potential of this value chain.

In covering a variety of issues, this book aspires to carve the path in the direction of the promotion and conservation of biodiversity through sustainable supply chains in the global agri-food sector.

We would like to thank Maria Verivaki, English professor, for the proofreading. We extend our appreciation to the authors and referrers of these chapters and Springer Academic Publications for their assistance in producing this book.

Thessaloniki, Greece — Konstadinos Mattas
Chania, Greece — George Baourakis
Chania, Greece — Constantin Zopounidis
Thessaloniki, Greece — Christos Staboulis

Contents

End-to-End Mapping Towards Sustainability: The Case of the Cretan Vegetable Supply Chain

Foivos Anastasiadis, Angelos Liontakis, and Giannis T. Tsoulfas

1 Introduction

One of the biggest problems that contemporary humankind is now experiencing is finding healthy and sustainable nourishment. Consumer concerns, which affect purchase decisions, are increasingly reflected in the quality of life and environmental preservation. Because crises and disruptions have historically detrimental consequences on food production, food distribution networks, and food consumption, purchasing behaviour can be put at risk during significant health emergencies (Brumă et al., 2021). The agri-food industry has seen a fast transformation over the past few decades as a result of several interrelated variables. Globalization, trade liberalization, population expansion, urbanization, rising incomes, changes in government regulations, dietary and food consumption trends, technology advancements, and environmental degradation are a few of these. Agri-food markets' functionality and structure, as well as their contribution to food and nutrition security and the sustainability of agri-food systems, have all been strongly impacted by specific trends formed by local and international forces (Borsellino et al., 2020).

The necessity for more sustainable practices in the agri-food sector is well documented, either in a more generic approach by proposing roadmaps (Ivo de Carvalho et al., 2022) and frameworks (Iakovou et al., 2014; Silvestri et al., 2022) or linked to specific concepts/needs such as traceability (Anastasiadis, 2022) and food security (Namany et al., 2020; Wahbeh et al., 2022). Despite the progress in the last few decades around this area, there is still room for improvement not only in developing economies but also in developed countries/markets. The case of the Greek agri-food supply chain is an interesting case study given its poor sustainability performance;

F. Anastasiadis (✉) · A. Liontakis · G. T. Tsoulfas
Department of Agribusiness and Supply Chain Management, Agricultural University of Athens, Athens, Greece
e-mail: foivos@aua.gr

K. Mattas et al. (eds.), *Value Chain Dynamics in a Biodiverse Environment*, Cooperative Management, https://doi.org/10.1007/978-3-031-49845-9_1

e.g. a study about the sustainable tomato in Greece highlighted a need for further interventions throughout the supply chain (Anastasiadis et al., 2020). Moreover, in Greece, there are several agri-food sectors (e.g. wine, livestock, fish) and regions with intense agribusiness development (e.g. Crete, the largest Greek island) that have not been studied sufficiently, supporting the necessity for further research.

The current work focuses on the Cretan vegetable supply chain and explores any possible intervention towards sustainable practices. The key driver is the lack of studies in that area combined with the importance of Crete in vegetable production. Furthermore, the complicated nature of the agri-food supply chains and the crucial role of the stakeholders involved (Tsoulfas et al., 2019), especially when the key concept under investigation is sustainability, dictate a holistic approach. Thus, end-to-end supply chain mapping was selected as the main methodology. In this way, all the stages, from the production to the final consumer, were considered, and every stakeholder participated in the study covering any interactions within the supply chain. The structure of this work involves a literature review of sustainability studies in the agri-food sector, highlighting the importance of the approach taken, as well as key papers that have employed mapping tools illustrating the value of the selected method. Then, we present methodological details of the mapping and specifics of the Cretan case study following the key results and discussion. Finally, there are the key conclusions, implications, and recommendations of the study.

2 Literature Review

2.1 Sustainability in the Agri-food Sector

Since a third of greenhouse gas emissions from human activities come from the food production system, issues such as climate change and ecosystem degradation influence people's overall health and well-being both directly and indirectly. The current system of food production is based on industrialized, massive agricultural output and long-distance supply chains. The distances that food travels and the number of middlemen between producers and consumers influence the environmental costs associated with packaging, transportation, conservation, and food waste. Furthermore, this type of arrangement creates stark inequalities between the parties involved, favouring merchants over farmers and distancing customers from the processes involved in food production (Craveiro et al., 2019). Concerns about how the value created by food systems is sustainably shared across supply chain participants in many nations and regions are raised by globalization. Increased globalization has been blamed for enormous supply chain disruptions. Even though globalization has led to the development of massive supply chains that have fuelled recent economic expansion, it has not sufficiently addressed the environmental, social, and political risks that have occasionally reduced supply efficiency (Nchanji & Lutomia, 2021). For many years, research and policy agendas have placed a high emphasis on sustainable development, techniques, and indicators for evaluating the existing condition and progress towards sustainability in the agri-food industry. A generalized dissatisfaction with the nutritional efficiency and sustainability of the modern, industrialized global food system pointed to the need for technological, social, and institutional advancements as well as potential alternative methods of food production, distribution, and consumption that could enable system change (Doernberg et al., 2022).

The circular economy is becoming more widely recognized as a means of promoting sustainable development, especially concerning environmental protection. This is especially the case for the agri-food sector, where a significant volume of organic waste and by-product streams provides a significant hurdle to many food processors (Klein et al., 2022). By moving towards circular food supply chains, food loss may be avoided, resource productivity can be increased, and regenerative natural systems can be created (Pakseresht et al., 2023).

Food markets have been affected by the growing consumer awareness of sustainable development. Consumption decisions are rapidly turning into a platform for people to express their values in opposition to mass-produced, internationally distributed food. Local agri-food products are becoming more and more popular because they are thought to be healthier, are unprocessed, and include fewer preservative ingredients. They also require fewer natural resources, which reduces their environmental impact (Barska & Wojciechowska-Solis, 2020; Kovács et al., 2022). From the perspective of the customer, the traditional worldwide food system, which involves enlarging supply chains for cost-effectiveness, has led to a distance between producers and consumers, frequently resulting in anonymity. Additionally, food crises and scandals have affected customers' faith in the food system as a whole and helped raise social knowledge of food-related facts (Corsi & Mazzocchi, 2019; Cruz et al., 2021).

In the academic literature, there have been proposed various frameworks that address sustainability in agri-food supply chains. Van der Vorst et al. (2013) provided a framework for sustainability research in the logistics of food supply chains, which included drivers, strategies, performance indicators, metrics, and possibilities for improvement. The conceptual framework proposed by Barth et al. (2017) emphasizes the necessity of including sustainability in business model innovation in the agri-food sector based on the study of the literature. Integrating sustainability elements into the three value chain building blocks—the value proposition, value creation and delivery, and value capture—will help managers create sustainable business models. Herrera-Reyes et al. (2018) examined the governance of an agri-food cooperative to suggest a strategy for the organization to include project management and project culture into its governance structure. It is predicated on the idea that project-based governance can help the agri-food system remain sustainable by assisting agri-food cooperatives in defining a governance structure that enables them to meet the demands of the current environment, accomplish their strategic goals, maintain their market share, and be sustainable. Silvestri et al. (2022) attempted to identify the most often discussed subjects concerning sustainability measurement indicators as well as how the indicators were applied to the three TBL pillars in the agri-food sector. They ended up with three clusters of analysis: LCA, best practices, and decision-making. Using an application to assess the water footprint of food consumption in Europe, Gibin et al. (2022) presented a new framework for the assessment of environmental impacts of food consumption patterns. This framework is intended to support large-scale assessments, which are primarily carried out by policymakers, and offers a standardized system-based method for analysing the environmental effects of food systems, measured in terms of their use of natural resources and their contribution to greenhouse gas emissions. Ivo de Carvalho et al. (2022) established a standard mapping of agri-food sustainable supply chains that will be taken into account by a performance measurement system, accurately identifying the major players, the direct and reverse flows of materials, information, and financial matters. Next, they determined which performance areas should be the most closely monitored in terms of sustainability. Finally, they described the key subcategories for each performance area that has been suggested.

As Greece is a small country, there are, as expected, only a few papers which provide evidence and empirical data on agri-food sustainability issues. Martzopoulou and Komninos (2019) focus on the issue of energy consumption in the agri-food sector. Greece was used as a case study to examine how energy sources affect the sustainability and environmental effects of the food industry. The diagnostics and benchmarking of the food industries' sustainability and environmental implications about the available energy source employed the energy analysis as a supportive tool. Trivellas et al. (2020) used a structured questionnaire which was sent to 134 executives of companies in the Greek agri-food sector. The results showed that the most significant influences on supply chain, business, and sustainable performance are information exchange, logistics networking, and transportation. Green packaging also has implications for economic and social performance. Anastasiadis et al. (2022) investigate the function of traceability systems in the shift to circular agri-food supply chains that are driven by sustainability in the Greek agri-food sector. They note that the adoption of circular economy practices has been slow and a coordinated strategy throughout the whole supply chain is necessary. Therefore, a holistic and multi-perspective strategy aligns supply chain stakeholders with final consumers as key actors and integrates management effectively. Karkanis and Melfou (2022) mainly dealt with the economic perspective of sustainable development in the Greek agri-food sector. In particular, they noted that the combined occurrence of the economic and pandemic crises has highlighted the pressing need for Greece's economy to re-evaluate sustainable economic growth plans. According to their results, increasing the exportability of Greek agri-food goods is a requirement for this industry's competitiveness in global markets. This may be accomplished by fostering synergies among agribusinesses and streamlining the standards-setting process for products. Chymis and Skouloudis (2015) provide a different approach. They assess the sustainable growth potential—mainly through a macroeconomic policy lens—in the agri-food sector in Greece by building on the lessons learnt in the case of New Zealand. Between 1984 and 1995, New Zealand had similar economic and political problems with Greece that necessitated significant changes to the way its development programme was constructed.

2.2 Mapping Vegetable Supply Chains

Supply chains are complicated systems with many activities, decision-making tasks, stakeholders, and interactions. To understand this complexity, a supply chain map—a visual depiction of all these components—is necessary (Anastasiadis & Alebaki, 2021). Supply chain mapping is crucial for business differentiation and constitutes a necessity that is becoming more and more critical. Businesses use it to achieve supply chains that are sustainable, integrated, and visible (Khan et al., 2022). By working with partners and other stakeholders, supply chain mapping may help an organization identify the opportunities and threats associated with its supply chain operations (Kashmanian, 2017).

In the sequel, several mapping examples from the agri-food sector are briefly presented to illustrate various applications of this approach. The Lao PDR's domestic and international supply networks for cabbage were investigated and mapped (Thongsavath et al., 2012). After that, a thorough study was carried out to gather information on postharvest procedures and estimate postharvest losses at various points in the local and export supply chains. Kumar et al. (2013) analysed food product supply networks, related industrial actors, and institutional support factors in the UK using a sector mapping technique. Publicly available industry reports, literature reviews, and case studies incorporating semi-structured interviews with important industry actors are some examples of data sources. A collection of maps outlining the organization of the UK food supply chains was created using an approach that involves a review of pertinent literature and government data. To enhance the application of lean manufacturing principles, Wesana et al. (2019) combined value stream mapping analysis with the food loss and waste methodology to pinpoint hotspots and assess the extent of food loss and waste along a Ugandan dairy value chain. Anastasiadis et al. (2020) presented three maps for the Greek tomato supply chain. The first one highlights the key elements that must be evaluated in a sustainability study by presenting the various processes and providing more information about their input-output. The tomato value chain map depicts the convoluted movement of goods and information while indicating possible critical points. The last map concentrates on the stakeholders and offers more information on the general comprehension of the sustainable tomato supply chain. Xue et al. (2021) plotted the mass flow and evaluated the effectiveness of the complete tomato farm-to-fork supply chain for the year 2016 using the European Union and its 28 member states as an example. Then, they looked at the possibilities of a wide range of GHG emissions mitigation measures, spanning from production efficiency improvement to process optimization, food waste reduction, trade pattern modification, and dietary structure change. Jayalath and Perera (2021) mapped Sri Lanka's supply chain and postharvest waste using value stream mapping and system dynamics. Value-adding and nonvalue-adding activities were distinguished using value stream mapping. A causal loop diagram was created using the information learned from the literature and industry inputs. El Hathat et al. (2023) elaborated on Morocco's food supply chain carbon emissions offsets. The study maps the food supply chain and agricultural production process to determine the environmental effect to achieve this goal. The study used machine learning-based prediction models to assess three distinct stages of tomato production in Morocco, including cultivation, harvesting, transport, and shipment. The investigation led to a strategy for carbon offsetting and provided insights into GHG emissions in tomato production cycles.

3 Materials and Methods

3.1 Supply Chain Mapping

There are a variety of tools available for mapping and modelling business processes or operations, as well as documenting the underlying material, information, and data flows (Van Der Aalst et al., 2004). In general, these frameworks establish and implement model-based principles that enable the conceptual representation of business processes in networks, which may then be used to guide quantitative analysis at different levels of abstraction (Aguilar-Savén, 2004).

This chapter uses the Business Process Model and Notation (BPMN) to capture the business processes, associated information, and material flows for the supply chains under exploration. The details about BPMN are specified by the Standard ISO/IEC 19510:2013 and the OMG Specification v2.0.2 (OMG, 2014). Rosing et al. (2015) as well as Chinosi and Trombetta (2012) summarize the basic BPMN formalisms. Appendix I contains a reference guide on the representations used in BPMN.

Essentially, BPMN enables the visual representation of business processes in a business process model by utilizing a uniform syntax (Teixeira & Borsato, 2019). The BPMN design requires that business processes be recorded in levels or layers. Open-source and commercial graphical tools are available to support the technique. The approach has been used for mapping purposes in a variety of areas, including (1) construction (Cheng et al., 2010), (2) food for traceability purposes (Pizzuti et al., 2014), and (3) healthcare for re-engineering processes (Wirachchaya & Duangpun, 2018).

BPMN was chosen as a mapping mechanism because it helps to maintain compatibility with recognized reference frameworks for the structured representation of business processes. Furthermore, BPMN aids in the high-level abstraction of business processes and underlying flows while ensuring that the same standard mapping concepts are implemented across the variety of crop supply chain systems studied in this work.

In the current work, the mapping was designed via the MS Visio Professional 2016, with a summary of the available tools that could be used in BPMN mapping offered by Geiger et al. (2018).

3.2 The Case of Cretan Vegetables

The aim of mapping the Cretan vegetable supply chain is to gain a deeper understanding and insights into the main features of the local agri-food system. The case study involves mapping the processes, crops, renewable feedstocks, actors, and value chains to reveal pathways for sustainability.

The mapping process involved two stages. In the first stage, a preliminary mapping of existing value chains was performed based on secondary evidence retrieved from the academic, institutional, and grey literature. Every map considers the system's boundaries and involved processes (e.g. farming inputs, crop varieties, processing stages, waste generation, retailing, and consumption), as well as value chain characteristics (e.g. types of inputs, valorization options, and outputs). The generated maps were then communicated to the particular stakeholders who possess the expertise and primary knowledge about each of the mapped supply networks. In the second stage, a validation process occurred and communication of the revised/validated map for final observations and comments took place. This two-stage process ensures a comprehensive and validated overview of the case under investigation.

Overall, by the end of August 2022, a total of two stakeholder involvement meetings had been organized. The engagement with the academic and industry informants from the local settings of the investigated case study helped to reduce bias and increase the reliability of the gathered data, thus informing the generated maps (Eisenhardt & Graebner, 2007). In total, 16 informants were involved, covering the entire supply chain, i.e. farmers, traders, scientists, retailers, and wholesalers. To ensure privacy, no personal identifying information of the informants is presented.

4 Results and Discussion

4.1 Initial (Theoretical) Map

The initial map (see Fig. 1) depicts the business processes as well as associated material and data flow for the Cretan vegetable supply chain that have been captured using secondary evidence. The map includes all the relevant processes that take place, specifically R&D and farm inputs supply, vegetable production plan, and retailing.

The upper part of the map describes the flows and processes in the inputs supply. It includes the R&D for developing new hybrid varieties based on specific needs such as increased yield and resistance to various diseases. In addition, it contains the processes needed to be approved by the corresponding public authorities to proceed with the commercial development and distribution of the new hybrid varieties to the plant nurseries.

The middle part of the map includes the processes that take part during the production phase of Cretan vegetables. It involves the design of the production (the

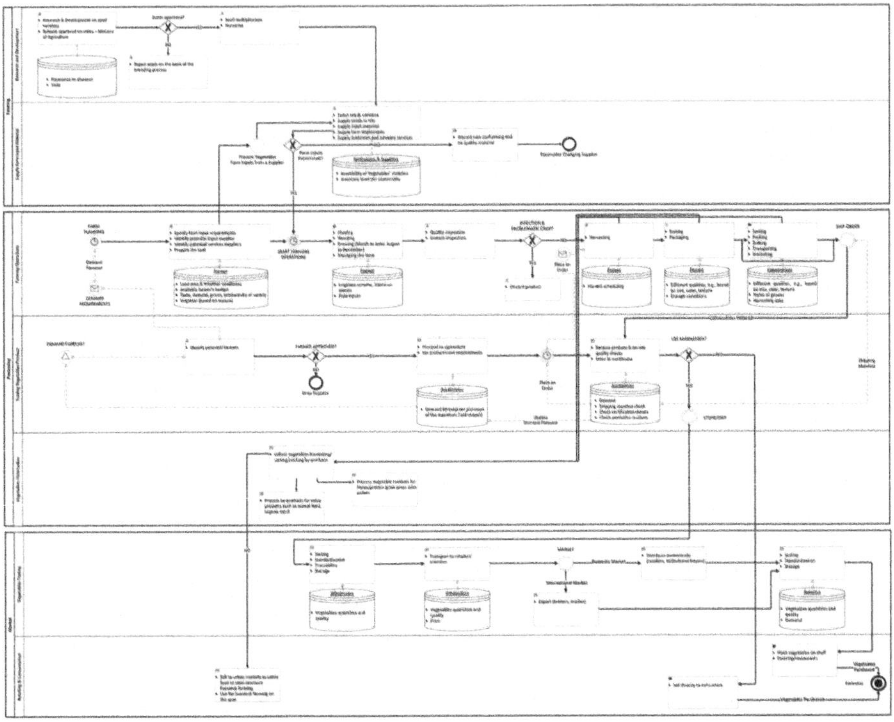

Fig. 1 Cretan fresh vegetables: Initial (theoretical) map

production plan), which is implemented by the farmers, mainly based on their perceived knowledge of the market, their budget constraints, and the expected weather conditions. Then, the production phase follows, ending with vegetable harvesting, based on a specific harvest plan. Finally, the produce is collected and distributed either individually or collectively to mediators (wholesalers, vegetable auctions) or directly to the consumers (through open markets). The part of the produce that cannot be directed to the market (either due to poor quality or low market prices) is headed for processing to produce fibre/protein grass, press cake pellets, or animal feed.

The final part of the map depicts the flows and processes in the vegetable market. It mainly consists of the flows and processes initiated by wholesalers that standardize, sort, and store the purchased vegetables before transporting them to retailers or other mediators in the national or international markets. The vegetables finally end up in retail stores or catering and restaurants. Finally, a smaller part of the produce, as was already mentioned in the description of the middle part of the map, ends up in the open markets (directly to the consumers).

4.2 Final (Validated) Map

The validated map is provided in Fig. 2. The validation procedure has proven to be very important as it reveals significant changes in the production flows and processes throughout the supply chain relative to the theoretical one. Interestingly, the validation does not provide any substantial change in the first (upper) part of the map that regards the supply of inputs. Therefore, the participatory approach did not alter the initial map designed based on secondary sources.

On the other hand, there are significant differences in the middle part of the validated map regarding the processes that take part during the production phase of Cretan vegetables. The growing phase starts a month earlier in Crete, especially in locations where the weather conditions are more appropriate for earlier production. In addition, quality certification is a critical element in several phases throughout the supply chain and is also a prerequisite for the distribution of vegetables in most locations. Moreover, quality certification is essential for traceability throughout the supply chain, which is the system used to trace back to any stage of the food supply chain to the origin of the vegetable produce (at the farm level). As traceability can reduce risks and losses due to possible product recalls, it is a very powerful

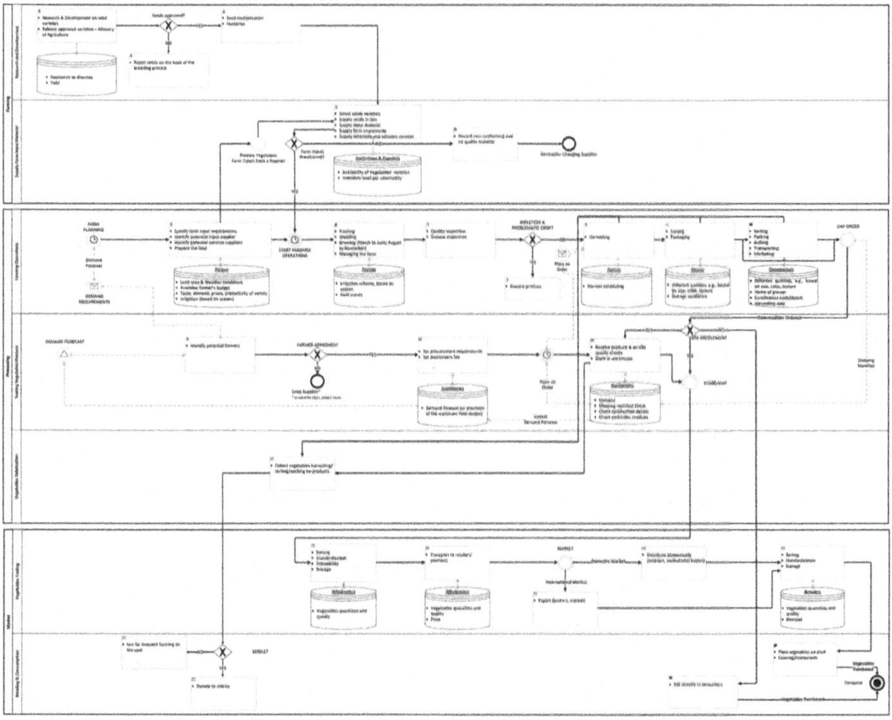

Fig. 2 Cretan fresh vegetables: Final (validated) map

tool and is intensively utilized by farmers' collective actions and (private or collective) auctions.

Another key element of the supply chain that came out throughout the validation process is the residue checks that various players in the supply chain, such as auction owners, wholesalers, and retailers, implement. Finally, the validation process reveals the limited control and utilization of the waste in the supply chain. Very few quantities are processed for animal feed production, while other potential procedures of circular economy (e.g. pellets production) are absent.

The key findings of the study concern a comprehensive understanding of the Cretan fresh vegetable supply chain, including all the processes and key stakeholders involved. Further analysis reveals areas of possible product and/or information bottlenecks across the entire supply chain, highlighting trading as a major issue for both producers and wholesalers. Despite the novelties and good practices, for example, in the auctions of Cretan fresh vegetables, there is still room for improvement. Vegetable growers require additional and constant training via extension agriculture programmes on production planning and sustainable farming practices. Another element underlined from the above analysis is the need for circular economy interventions regarding non-market farm produce. Further work towards this direction must be done by the key actors of the supply chain. Initiations provided by the local authorities may also work towards this direction. Certification schemes and traceability systems are considered key elements for a healthy-operated and sustainable vegetable supply chain with smooth flows and processes.

5 Conclusions

The current work provides a comprehensive outlook of the Cretan fresh vegetable supply chain. Employing an end-to-end supply chain mapping ensured that the analysis captured every process, stakeholder, and interaction. Such a holistic approach to visualizing the bigger picture that also provides the ability to reveal specific areas that require further exploration is the key contribution of this study.

Specifically, the findings suggest critical interventions covering sustainability's triple bottom line. Concerning the economic pillar, adopting digital tools in trading could result in better prices for both traders and producers. Furthermore, utilizing historical sales data at a regional level could result in demand forecasting and evidence-based guidelines for future production, e.g. best-selling varieties. Regarding the environmental pillar, as a response to increasing consumer demand for sustainably produced food products, adopting more environmentally friendly practices across the supply chain is essential. Finally, as far as the social pillar is concerned, organized actions towards food waste reduction throughout the supply chain, i.e. production, sorting, standardizing, and trading by donating significant food quantities to charity (circular economy), can enhance the social sustainability level. Moreover, the social cohesion among the local population can be enhanced and the prospects for local and regional development can be improved.

There are several implications and recommendations based on the above findings. As this study offers a thorough end-to-end understanding of the Greek fresh vegetable supply chain, further studies may utilize this to design critical interventions to improve the performance of the entire supply chain. The study also has some other practical implications related to environmental issues consistent with the response to consumer pressure to prove environmental credentials. Agri-food companies are always trying to provide a better image for the public by being environmentally friendly. For example, employing a traceability system for protecting environmental resources can become a primary tool for enhancing trust and transparency.

Additionally, it is important to mention that contributing towards sustainable practices also enhances the adoption of a circular economy rationale, thus improving the overall supply chain performance and environmental footprint, e.g. reducing recalls/food waste and reusing packages or even food waste in the supply chain. Moreover, carefully designed and implemented policy initiatives can prove useful for a more sustainable supply chain that utilizes circular economy schemes. Finally, stakeholder involvement is crucial to validate and fine-tune a supply chain map, as it is a means of collecting non-reported knowledge. Therefore, enhancing stakeholder participation can add value to the research outcomes and assist in an efficient and to-the-point co-creation of policy design for a more sustainable supply chain.

Acknowledgement The authors would like to acknowledge the Unified Agricultural Association of Ierapetra for their assistance throughout the project, as well as Prof. Konstantinos Karantininis, Dr. Alexandra Sintori, and Dr. Antonios Proestakis for their contribution to the data collection and validation.

FundingThis research is part of the Programme Agreement between Region of Crete and the Agricultural University of Athens entitled "Investigating the benefits of a unified digital vegetable auction in Ierapetra" and financed by the Region of Crete.

Appendix 1: BPMN Constructs and Symbols

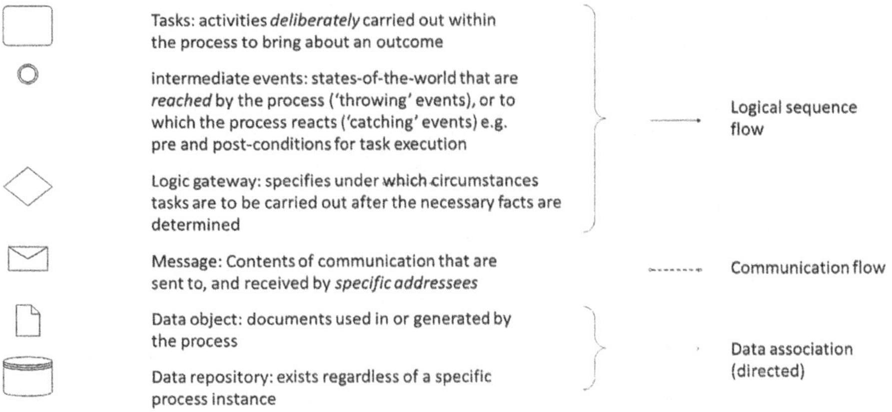

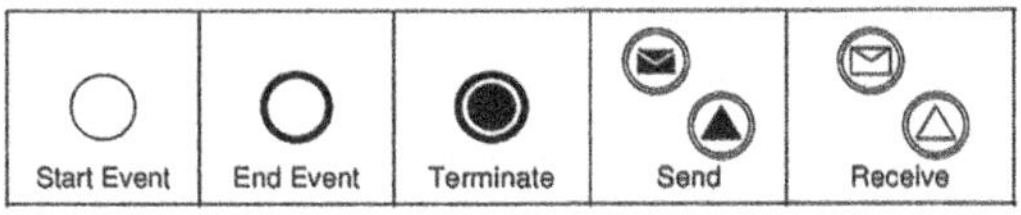

References

Aguilar-Savén, R. S. (2004). Business process modelling: Review and framework. *International Journal of Production Economics, 90*, 129–149.

Anastasiadis, F. (2022). Designing a traceability framework for sustainable agri-food supply chains. In K. Mattas, G. Baourakis, C. Zopounidis, & C. Staboulis (Eds.), *Food policy modelling: Responses to current issues*. Springer International Publishing.

Anastasiadis, F., & Alebaki, M. (2021). Mapping the Greek wine supply chain: A proposed research framework. *Foods, 10*, 2859.

Anastasiadis, F., Apostolidou, I., & Michailidis, A. (2020). Mapping sustainable tomato supply chain in Greece: A framework for research. *Foods, 9*.

Anastasiadis, F., Manikas, I., Apostolidou, I., & Wahbeh, S. (2022). The role of traceability in end-to-end circular agri-food supply chains. *Industrial Marketing Management, 104*, 196–211.

Barska, A., & Wojciechowska-Solis, J. (2020). E-consumers and local food products: A perspective for developing online shopping for local goods in Poland. *Sustainability, 12*.

Barth, H., Ulvenblad, P.-O., & Ulvenblad, P. (2017). Towards a conceptual framework of sustainable business model innovation in the agri-food sector: A systematic literature review. *Sustainability, 9*.

Borsellino, V., Schimmenti, E., & El Bilali, H. (2020). Agri-food markets towards sustainable patterns. *Sustainability, 12*.

Brumă, I. S., Vasiliu, C. D., Rodino, S., Butu, M., Tanasă, L., Doboş, S., Butu, A., Coca, O., & Stefan, G. (2021). The behavior of dairy consumers in short food supply chains during COVID-19 pandemic in Suceava Area, Romania. *Sustainability, 13*.

Cheng, J. C. P., Law, K. H., Bjornsson, H., Jones, A., & Sriram, R. D. (2010). Modeling and monitoring of construction supply chains. *Advanced Engineering Informatics, 24*, 435–455.
Chinosi, M., & Trombetta, A. (2012). BPMN: An introduction to the standard. *Computer Standards & Interfaces, 34*, 124–134.
Chymis, A., & Skouloudis, A. (2015). Far away, so close? Examining the growth potential of Greece through the lens of New Zealand's paradigm. In M. Di Bitetto, A. Chymis, & P. D'anselmi (Eds.), *Public management as corporate social responsibility: The economic bottom line of government*. Springer International Publishing.
Corsi, S., & Mazzocchi, C. (2019). Alternative food networks (AFNs): Determinants for consumer and farmer participation in Lombardy, Italy. *Agricultural Economics (Czech Republic), 65*, 259–269.
Craveiro, D., Marques, S., Marreiros, A., Bell, R., Khan, M., Godinho, C., Quiroga, S., & Suárez, C. (2019). Equity, health, and sustainability with prove: The evaluation of a Portuguese program for a short distance supply chain of fruits and vegetables. *International Journal of Environmental Research and Public Health, 16*.
Cruz, J. L., Puigdueta, I., Sanz-Cobeña, A., & González-Azcárate, M. (2021). Short food supply chains: Rebuilding consumers' trust. *New Medit, 20*, 33–47.
Doernberg, A., Piorr, A., Zasada, I., Wascher, D., & Schmutz, U. (2022). Sustainability assessment of short food supply chains (SFSC): Developing and testing a rapid assessment tool in one African and three European city regions. *Agriculture and Human Values, 39*, 885–904.
Eisenhardt, K. M., & Graebner, M. E. (2007). Theory building from cases: Opportunities and challenges. *Academy of Management Journal, 50*, 25–32.
El Hathat, Z., Sreedharan, V. R., Venkatesh, V. G., Zouadi, T., Arunmozhi, M., & Shi, Y. (2023). Modelling and analyzing the GHG emissions in the VUCA world: Evidence from tomato production in Morocco. *Journal of Cleaner Production, 382*, 134862.
Geiger, M., Harrer, S., Lenhard, J., & Wirtz, G. (2018). BPMN 2.0: The state of support and implementation. *Future Generation Computer Systems, 80*, 250–262.
Gibin, D., Simonetto, A., Zanini, B., & Gilioli, G. (2022). A framework assessing the footprints of food consumption. An application on water footprint in Europe. *Environmental Impact Assessment Review, 93*, 106735.
Herrera-Reyes, A. T., Carmenado, I. D., & Martínez-Almela, J. (2018). Project-based governance framework for an agri-food cooperative. *Sustainability, 10*.
Iakovou, E., Vlachos, D., Achillas, C., & Anastasiadis, F. (2014). Design of sustainable supply chains for the agrifood sector: A holistic research framework. *Agricultural Engineering International: CIGR Journal*. Special issue 2014: Agri-food and biomass supply chains, 1–10.
Ivo De Carvalho, M., Relvas, S., & Barbosa-Póvoa, A. P. (2022). A roadmap for sustainability performance assessment in the context of Agri-Food Supply Chain. *Sustainable Production and Consumption, 34*, 565–585.
Jayalath, M. M., & Perera, H. N. (2021). Mapping post-harvest waste in perishable supply chains through system dynamics: A Sri Lankan case study. *Journal of Agricultural Sciences – Sri Lanka, 16*, 526.
Karkanis, D., & Melfou, K. (2022). Reconsidering the Greek agri-food export sector: What lessons from the EU partners? *Global Business Review*, 09721509221135928.
Kashmanian, R. M. (2017). Building greater transparency in supply chains to advance sustainability. *Environmental Quality Management, 26*, 73–104.
Khan, S. A., Mubarik, M. S., Kusi-Sarpong, S., Gupta, H., Zaman, S. I., & Mubarik, M. (2022). Blockchain technologies as enablers of supply chain mapping for sustainable supply chains. *Business Strategy and the Environment, 31*, 3742–3756.
Klein, O., Nier, S., & Tamásy, C. (2022). Circular agri-food economies: Business models and practices in the potato industry. *Sustainability Science, 17*, 2237–2252.
Kovács, I., Lendvai, M. B., & Beke, J. (2022). The importance of food attributes and motivational factors for purchasing local food products: Segmentation of young local food consumers in Hungary. *Sustainability, 14*.

Kumar, M., Srai, J., Pattinson, L., & Gregory, M. (2013). Mapping of the UK food supply chains: Capturing trends and structural changes. *Journal of Advances in Management Research, 10*, 299–326.

Martzopoulou, A., & Komninos, N. (2019). The effect of solar energy on the environmental impacts and sustainability of food industry. *Journal of Environmental Accounting and Management, 7*, 303–321.

Namany, S., Govindan, R., Alfagih, L., Mckay, G., & Al-Ansari, T. (2020). Sustainable food security decision-making: An agent-based modelling approach. *Journal of Cleaner Production, 255*, 120296.

Nchanji, E. B., & Lutomia, C. K. (2021). COVID-19 challenges to sustainable food production and consumption: Future lessons for food systems in eastern and southern Africa from a gender lens. *Sustainable Production and Consumption, 27*, 2208–2220.

Pakseresht, A., Yavari, A., Kaliji, S. A., & Hakelius, K. (2023). The intersection of blockchain technology and circular economy in the agri-food sector11. This work was supported by the Swedish University of Agricultural Sciences, Sweden, by a scholarship from L Nannesson's foundation, grant number Dnr SLU.ua.2019.3.1.5-617. *Sustainable Production and Consumption, 35*, 260–274.

Pizzuti, T., Mirabelli, G., Sanz-Bobi, M. A., & Goméz-Gonzaléz, F. (2014). Food Track & Trace ontology for helping the food traceability control. *Journal of Food Engineering, 120*, 17–30.

Silvestri, C., Silvestri, L., Piccarozzi, M., & Ruggieri, A. (2022). Toward a framework for selecting indicators of measuring sustainability and circular economy in the agri-food sector: A systematic literature review. *The International Journal of Life Cycle Assessment*.

Teixeira, K. C., & Borsato, M. (2019). Development of a model for the dynamic formation of supplier networks. *Journal of Industrial Information Integration, 15*, 161–173.

Thongsavath, C., Varit, S., Thananya, W., Sirichai, K., & Antonio, A. (2012). Cabbage supply chain mapping and postharvest loss in the domestic and export trade in Lao PDR. *International Food Research Journal, 19*, 1615–1620.

Trivellas, P., Malindretos, G., & Reklitis, P. (2020). Implications of green logistics management on sustainable business and supply chain performance: Evidence from a survey in the Greek agri-food sector. *Sustainability, 12*.

Tsoulfas, G. T., Marinagi, C., Trivellas, P. & Reklitis, P. (2019). Sustainable agri-food supply chains in Greece: The role of stakeholders. In: ICMEF (Ed.) *2nd International Conference on Management, Economics & Finance*. Rotterdam.

Van Der Aalst, W., Van Hee, K. M., & Van Hee, K. (2004). *Workflow management: Models, methods, and systems*. MIT Press.

Van Der Vorst, J. G. A. J., Peeters, L., & Bloemhof, J. M. (2013). Sustainability assessment framework for food supply chain logistics: Empirical findings from Dutch food industry. *International Journal on Food System Dynamics, 4*, 130–139.

Von Rosing, M., White, S., Cummins, F., & De Man, H. (2015). Business process model and notation—BPMN. In M. Von Rosing, A.-W. Scheer, & H. Von Scheel (Eds.), *The complete business process handbook*. Morgan Kaufmann.

Wahbeh, S., Anastasiadis, F., Sundarakani, B., & Manikas, I. (2022). Exploration of food security challenges towards more sustainable food production: A systematic literature review of the major drivers and policies. *Foods [Online], 11*.

Wesana, J., Gellynck, X., Dora, M. K., Pearce, D., & De Steur, H. (2019). Measuring food losses in the supply chain through value stream mapping: A case study in the dairy sector. In C. M. Galanakis (Ed.), *Saving food*. Academic Press.

Wirachchaya, C., & Duangpun, K. (2018). A process reference model for hospital supply chain of pharmaceutical products. *Industrial Engineering & Management Systems, 17*, 43–61.

Xue, L., Cao, Z., Scherhaufer, S., Östergren, K., Cheng, S., & Liu, G. (2021). Mapping the EU tomato supply chain from farm to fork for greenhouse gas emission mitigation strategies. *Journal of Industrial Ecology, 25*, 377–389.

Agri-food Trade Competitiveness in Latin America and the Caribbean Region

Jeremiás Máté Balogh

1 Introduction

Latin America and the Caribbean (LAC) represent an extensive region from the Bahamas and Mexico to Brazil, Argentina, and Chile. The LAC region has great potential to significantly increase the competitiveness of the agri-food trade and become a leader in the global market. However, this region faces several challenges that hinder its competitiveness. Therefore, it is necessary to assess LAC's competitiveness in the agri-food sector. Despite the global relevance of the topic, research on the patterns and dynamics in Latin America and the Caribbean is scarce compared to other regions of the world. The chapter provides an overview of trade competitiveness and revealed comparative advantages in the agri-food sector, focusing on the LAC region. The chapter is organised as follows. The second section presents recent statistics on the agriculture of the LAC region. The third section discusses the importance of the Latin American and Caribbean economies, which justifies the relevance of the topic. The fourth section provides a review of the literature on competitiveness and revealed comparative advantages in the agri-food sector and identifies research gaps in the field. The fifth section introduces the measurement of comparative advantages in agricultural trade, applying Balassa indices and their modified versions. The third section covers competitiveness and revealed comparative advantages in the agri-food sector investigating the LAC region. The following section discusses empirical studies and their findings on LAC agricultural competitiveness. Cooperation as a potential solution to agricultural competitiveness is then discussed. Finally, the last section summarises the chapter and concludes.

J. M. Balogh (✉)
Department of Agricultural Economics, Institute of Sustainable Development, Corvinus University of Budapest, Budapest, Hungary
e-mail: jeremias.balogh@uni-corvinus.hu

K. Mattas et al. (eds.), *Value Chain Dynamics in a Biodiverse Environment*, Cooperative Management, https://doi.org/10.1007/978-3-031-49845-9_2

2 Recent Trends in Agricultural Development in the LAC Region

The agricultural sector represents up to 20% of the value added to the gross domestic product (GDP) in Latin American and Caribbean countries (Statista, 2023a). In this region, the agricultural sector[1] in Haiti contributed with 20.4% of the value added to the country's GDP, which is the highest contribution reported for the Latin America and Caribbean region in 2020. In contrast, the lowest contributions were registered by Puerto Rico and the Bahamas, with a share of less than 1% (Statista, 2023b). Regarding agricultural employment in LAC, the agricultural sector in Guatemala represented more than 31% of total employment in the country (2020), making it the LAC nation with the highest share of employment in the sector. This industry contributed with 30.6% of the country's total employment that year in Nicaragua, while in the case of Argentina, this share added up to less than 1% (Statista, 2023c). Taking into account performance, competitiveness, real GDP growth, and human development in the Latin America and Caribbean region, Reyes and Useche (2019) show that decreasing conditions in competitiveness and economic growth indicators are the illustrative situation in this region since 2009. The most competitive country in the region is Chile, while the weakest is Venezuela. The next section points out the importance of Latin American and Caribbean trade.

3 Importance of Latin American and Caribbean Trade

In 2020, the Latin America and Caribbean region had a total export of more than USD931 billion and imports of USD873 billion, leading to a positive trade balance of USD58 billion. Exports of goods and services as a percentage of GDP in the LAC region were 23.69%, while imports of goods and services as a percentage of GDP were 22.65% (World Bank, 2023a). Recent statistics have shown that Latin American and Caribbean exports continued to grow at a slower pace in 2022 than in the same period of the previous year. After growing 27.9% in 2021, the value of the region's exports increased by 20.6% in the first half of 2022. This growth rate exceeds that of world trade, which fell from 25.8% to 17.5% in the same period. Exports slowed in response to a series of global shocks, such as the Russian-Ukrainian conflict, the zero-COVID policy of China, and the tightening of monetary policies (Inter-American Development Bank, 2022). The value of Latin American and Caribbean food products represented USD70.9 billion with a 7.6% share in total LAC exports in 2020. The main destinations for LAC agricultural products were regions of North America and the region of Europe and Central Asia. The

[1] The agricultural sector includes forestry, hunting, and fishing, as well as cultivation of crops and livestock.

United States was the most important food export destination, followed by China, the Netherlands, Vietnam, Spain, and the United Kingdom in 2020 (Table 1).

Regarding trade, exports of food products increased significantly, especially since 2003. At the same time, the export share of these products varied between 6% and 8% between 2005 and 2019 (Fig. 1).

These statistics show the growing importance of agro-food products in LAC. Over the last 10 years, exports from Latin America and the Caribbean have been less dynamic and more volatile. In addition, the Inter-American Development Bank (2022) concludes that the recurring global shocks that have affected the regional trade performance point to a trend of instability in the medium term. Therefore, it is important to understand how competitive the LAC agricultural sectors are and how their comparative advantages are changed. The following section discusses potential definitions and the measurement of competitiveness and comparative advantages.

4 Definition and Measurement of Competitiveness and Comparative Advantages

Competitiveness and comparative advantage are complex notions, having many different definitions and approaches that can be interpreted at the micro (firm), mezzo (industry), or macro (national) level. At the macroeconomic level, competitiveness or comparative advantages are not well defined; there is a lack of a commonly accepted definition in economics. According to the literature, macro-level competitiveness is significantly linked to international trade performance. The definition of trade competitiveness provided by Jámbor and Babu (2016) is associated with the notion of comparative advantage, which means an economy's ability to produce

Table 1 Leading export destination countries for LAC food products in 2020

Destination country	Export in thousand USD	Export product share (%)
United States	19,967,261	4.87
China	3,406,443	2.63
Netherlands	2,884,289	16.72
Vietnam	1,826,711	31.79
Spain	1,583,945	14.97
United Kingdom	1,494,338	17.73
South Korea	1,494,268	9.55
Chile	1,355,651	11.13
Belgium	1,241,917	18.07
Germany	1,224,776	8.29
Rest of the World	35,574,374	2.0
World total	70,829,203	7.6

Source: World Bank (2023b)

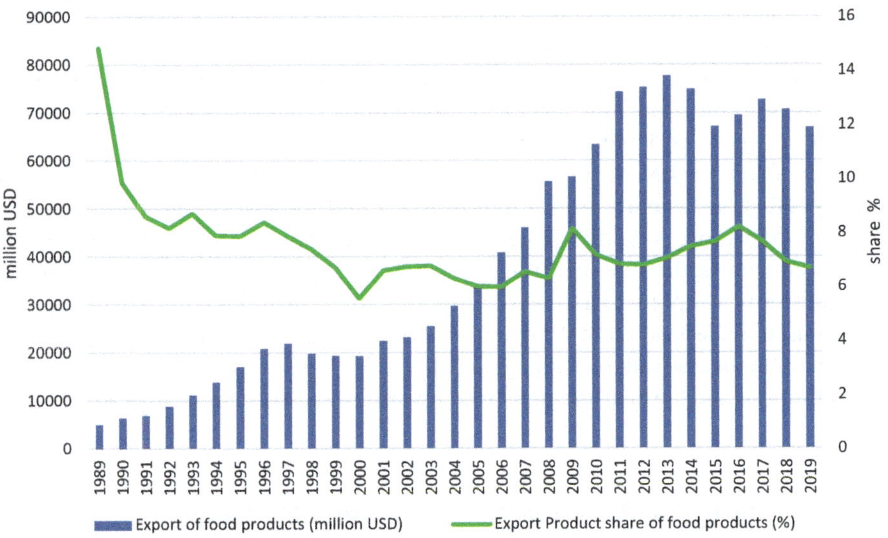

Fig. 1 Evolution of exports and export share of food products in LAC, 1989–2019. Source: World Bank (2023b)

goods or services at a lower opportunity cost than its trading partners or competitors. Competitive performance can be determined by different measurements based on indicators such as revealed comparative advantage, market performance, trade success, and others (Latruffe, 2010). Revealed comparative advantage (RCA) indices are the most widely used indicators for the trade relation competitiveness of nations. This method was formulated by Balassa (1965), a Hungarian economist who lived and worked in the United States. He calculated the ratio of a country's export share of a commodity on the international market to the exports of all commodities compared to the similar share of a group of countries or regions. The Balassa index was further modified by several authors, for example: Vollrath (1991), Dalum et al. (1998), Hoen and Oosterhaven (2006), and Yu et al. (2009). In the following sections, a wide range of indicators are presented to analyse the revealed comparative advantage for trade. These indicators are mainly based on the index developed by Balassa (RCA) and its latest developments (RTA, SRCA, RC, and NRCA), which can be calculated to identify the advantages or disadvantages by country, region, and sector.

5 Indices Capturing Revealed Comparative Advantages and Trade Competitiveness

To interpret and measure trade competitiveness or comparative advantages, the index of revealed comparative advantage (RCA) can be applied (Balassa, 1965). This measures the proportion of a country's exports for a single commodity to the exports of all commodities, and a similar share for a set of selected countries (Bojnec & Fertő, 2017). This index is widely used to evaluate agricultural competitiveness at the macro/national level. The revealed comparative advantage index is defined as follows (Liesner, 1958; Balassa, 1965):

$$RCA_{ij} = \left(\frac{X_{ij}}{X_{it}}\right) / \left(\frac{X_{nj}}{X_{nt}}\right) \tag{1}$$

where:

X represents exports.
i is a country.
j is a commodity.
t is a set of commodities.
n is a set of countries that are used as benchmark export markets for comparisons.

The index measures a country's exports of a commodity relative to its total exports and the corresponding export performance of a set of countries, for example, global agri-food exports. If B > 1, a country's comparative export advantage is revealed on the global market, while if B is between 0 and 1, it indicates a comparative disadvantage in agri-food trade. Despite its popularity, the RCA has faced several criticisms, particularly for disregarding the impacts of agricultural policy and other economic interventions, which can lead to an overestimation of comparative advantages. Therefore, the RCA calculation is based on export statistics, because export is less exposed to customs, tariffs, and agricultural policy distortions. Furthermore, the indicator is condemned for not providing asymmetric values because its advantages vary from 1 to infinite. If the index ranges from 0 to 1, the selected country has a comparative disadvantage, while when the index exceeds the value of 1, the country has a comparative advantage of overestimating the relative weight of a sector (De Benedictis et al., 2004; Jámbor & Babu, 2016; Bojnec and Fertő, 2019; Mizik et al., 2020). Building on the concept of Balassa, Vollrath (1991) proposed three distinct specifications of revealed comparative advantage to overcome the shortcomings of the Balassa index. First, he suggests the relative import advantage index (RMA), which is the counterpart of the RCA index (1) that captures import advantages (Vollrath, 1991):

$$RMA_{ij} = \left(\frac{M_{ij}}{M_{it}}\right) / \left(\frac{M_{nj}}{M_{nt}}\right) \tag{2}$$

where:

M denotes the imports.
i is a country.
j is a commodity.
t is a set of commodities.
n is a set of countries.

The positive values of RMA indicate revealed import advantages, while the negative values show the disadvantages. In addition to this, Vollrath (1991) proposed a different specification of revealed comparative advantage, called the relative trade advantage (RTA), which comprises the export and import side of the Balassa indices (RCA and RMA). The RTA is calculated as follows:

$$\mathrm{RTA}_{ij} = \mathrm{RCA}_{ij} - \mathrm{RMA}_{ij} = \left[\frac{\left(\mathrm{X}_{ij} / \mathrm{X}_{it}\right)}{\left(\mathrm{X}_{nj} / \mathrm{X}_{nt}\right)}\right] - \left[\frac{\left(\mathrm{M}_{ij} / \mathrm{M}_{it}\right)}{\left(\mathrm{M}_{nj} / \mathrm{M}_{nt}\right)}\right] \tag{3}$$

where:

X denotes the exports.
M denotes imports.
i is a country.
j is a commodity.
t is a set of commodities.
n is a set of countries.

In the case of RTA > 0, a relative trade advantage is revealed and the analysed country's trade is relatively more competitive in a given sector. Vollrath's (1991) third approach calculated the natural logarithm of the RCA and RMA indices and measured the difference between them, resulting in the revealed competitiveness index (RC) which reveals competitiveness for positive values and non-competitiveness for negative values:

$$\mathrm{RC}_{ij} = \ln\left(\mathrm{RCA}_{ij}\right) - \ln\left(\mathrm{RMA}_{ij}\right) \tag{4}$$

Furthermore, Dalum et al. (1998) developed an innovative method for dealing with the asymmetric problem of the RCA and constructed the revealed symmetric comparative advantage (RSCA) index:

$$\mathrm{RSCA} = \frac{\left(\mathrm{RCA} - 1\right)}{\left(\mathrm{RCA} + 1\right)} \tag{5}$$

Values between 0 and +1 for RSCA assume that the country has a comparative export advantage, while values −1 and 0 suggest a comparative export disadvantage. RSCA avoids the problem of zero values, which occurs in the logarithmic transformation (4) when an arbitrary constant is not added to the RCA. As the

SRCA distribution is symmetric around the origin, a possible bias is eliminated using this index (Dalum et al., 1998). This tool is also a useful alternative to receive indices close to the normal distribution. The stability and specialisation of RSCA can be tested using the following regression equation, country by country:

$$\mathrm{RSCA}_{ij}^{t2} = \alpha_i + \beta_i \mathrm{RSCA}_{ij}^{t1} + \varepsilon_{ij} \quad (6)$$

In the equation above, t_1 and t_2 refer to the initial year and the final year, respectively. The dependent variable of RSCA at time t_2 for sector i is tested against the independent variable of RSCA, which is the value of RSCA in the previous year t_1. α and β are the parameters of standard linear regression, and ε denotes the estimated residuals. To eliminate the problems of the asymmetric nature of the RCA index, Hoen and Oosterhaven (2006) presented an innovative technique of revealed comparative advantage (ARCA):

$$\mathrm{ARCA}_{ij} = \left(\frac{\mathrm{X}_{ij}}{\mathrm{X}_{it}}\right) - \left(\frac{\mathrm{X}_{nj}}{\mathrm{X}_{nt}}\right) \quad (7)$$

where:

X denotes the exports.
M denotes imports.
i is a country.
j is a commodity.
t is a set of commodities.
n is a set of countries.

The ARCA is a symmetric index that ranges from −1 to +1 with a turning point of 0. Furthermore, Yu et al. (2009) invented a normalised revealed comparative advantage (NRCA) index to calculate the dynamics of comparative advantage and to improve certain features of the original index (RCA) in static patterns in comparative advantage, as an appropriate export specialisation index for comparison in comparative advantage. Yu et al. (2009) defined the NRCA as follows:

$$\mathrm{NRCA}_{ij} = \left(\frac{E_{ij}}{E}\right) - \left[\left(\frac{E_i}{E}\right) / \left(\frac{E_j}{E}\right)\right] \quad (8)$$

where:

E denotes total world trade.
E_{ij} describes the actual export of commodity j on the world market.
E_i is the export of all commodities.
Ej denotes the export of the commodity j by all countries.

NRCA > 0 suggests that the comparative advantage of a country on the world market is revealed, while NRCA < 0 signifies disadvantages. The distribution of the

NRCA is symmetric, ranging from −1/4 to +1/4, with 0 being a neutral point of comparative advantage. Among others, Lafay (1992) proposed a more comprehensive index with greater explanatory power. Compared to the RCA, the advantage of the Lafay index is the ability to include any distortions caused by macroeconomic instabilities. This index allows for analysing the position of every specific product within the foreign trade structure of every analysed country or a group of countries. The Lafay index is defined as follows (Lafay, 1992):

$$LI_{ij} = 100\left[\frac{(X_{ij} - M_{ij})}{(X_{ij} + M_{ij})} - \frac{\sum_k (X_{ik} - M_{ik})}{\sum_k (X_{ik} + M_{ik})}\right]\left(\frac{(X_{ij} + M_{ij})}{\sum_k (X_{ik} + M_{ik})}\right) \tag{9}$$

where:

X represents exports.
M represents imports.
i is the country.
j is a given product.
k is the number of items analysed.

X_{ij} and M_{ij} represent exports or imports of product *j* realised by country *i* or a group of countries *i* with respect to the rest of the world or with respect to a selected business partner (partner country). The index ranges from − to +. If its value is higher (or lower) than zero, the country has (or doesn't have) a comparative advantage on the world market in exporting products from a selected industry. As this indicator takes into account both exports and imports, the index is more suitable for intra-industry trade analysis. The Lafay (1992) index is directly related to the Grubel–Lloyd index of intra-industry trade. Comparing the results obtained by using the Balassa and Lafay indices shows that they generate a radically different representation of the product space (Ferrarini & Scaramozzino, 2011).

It should be mentioned that the methodology described above has several limitations. First, one of the most significant issues is the complexity of the world food trade system. Today, trade takes place at all levels (firm, country, or region). Furthermore, since agricultural commodities are essential for food security, their trading system is very intense, making it difficult to aggregate and record the exact quantity and value of agricultural products traded. As a result, trade values may not always count in the total trade value of a particular country. Second, another challenge arises when there are missing observations in trade data, e.g. when two nations do not trade with each other or the amount of trade is so small that it is not recorded. This can lead to under- or overestimation of trade indices. Additionally, each index has its own set of constraints, such as measuring the export or import side, asymmetry, policy-induced distortions, and market intervention. While RTA and RC incorporate RMA taking into account the import side of trade in their calculation method, they are more likely to be impacted by policy and governmental interventions; the original RCA, SRCA, and NRCA neglect import advantages (Török & Jámbor, 2016). Finally, French (2017) concluded that RCA indices are usually not reliable with theoretical models of comparative advantage, but they are suitable when exposing the fundamental patterns of comparative advantage, evaluating the differential effect of changes in trade barriers between different products, or identifying countries that are competitors in a particular market.

6 Competitiveness and Comparative Advantages in Different Agricultural Sectors

Many studies have been published on analysing comparative advantages in different agricultural sectors (Mizik, 2021 identified more than 50). The aim of this section is not to summarise all of them but to highlight the most important ones and to summarise the findings relevant to agricultural trade. Generally, certain authors that publish the most frequently on the topic of comparative advantages in agricultural sectors include Bojnec, Fertő, Jámbor, Mizik, Balogh, and Török. The first group of studies available in the economics literature analysed the competitiveness of agricultural trade at the global level (Table 2).

6.1 Studies at the Global Level

In their book, Jámbor and Babu (2016) examined the competitiveness of global agriculture by calculating the revealed comparative advantage (RCA) index for all countries and agricultural products for the period 1991–2014. The scholars calculated an average of all periods analysed and concluded that the most competitive nations are highly developed agricultural producers such as the Netherlands, Spain, and Denmark, while small states such as Montserrat, Brunei, and the Cook Islands were the least competitive, demonstrating comparative disadvantage. Some years later, Jámbor et al. (2018) analysed global spice trade competitiveness by examining Balassa's RCA from 1991 to 2015. They observed that the market was concentrated in Guatemala, Sri Lanka, and India, which had the highest RCA indices during the period; on the other hand, despite being the largest exporters, Germany and the Netherlands, as the main processors and sellers of spices, had a comparative disadvantage in the world's spice trade. The second most important group were the regional and country-specific articles that revealed the competitiveness of different individual countries that used RCA and more advanced NRCA indices.

Table 2 Research on comparative advantages in different agricultural sectors

Author(s)	Products	Indicator	Period	Region
Global level				
Jámbor and Babu (2016)	Agriculture	RCA	1991–2014	Global level
Jámbor et al. (2018)	Spice trade	RCA	1991–2015	Global level
Regional level				
Bojnec (2001)	Agricultural trade	RCA, RTA, Grubel–Lloyd index (GL) Terms of trade (TT)	1992–1997	Central and East Europe European Union
Fertő (2008)	Agri-food trade	RCA, persistence, and mobility	1992–2002	Central European countries
Török and Jámbor (2016)	Ham trade	RCA, RSCA Kaplan–Meier estimator for survival function	1999–2013	European Union
Bojnec and Fertő (2018)	Agri-food export	NRCA	2000–2011	European Union
Mizik et al. (2020)	Agri-food trade	RCA stability and duration	2000–2015	Commonwealth of Independent States
Hoang (2020)	Agricultural products: rice, rubber, spices, vegetable fats and oils, wood, fuel wood, fish, and crustaceans	RCA, RTA, and NRCA	1997–2015	ASEAN
Country-specific studies				
Fertő and Hubbard (2003)	Agriculture and food processing	RCA, RMA, RTA, RC	1992–1998	Hungary
Bakhshinejad and Hassanzadeh (2012)	Agricultural products: walnut, almond hazelnut, apple, orange	RCA	2007–2002	Iran
Burianová and Belová (2012)	Agricultural trade, 24 commodity chapters	RCA, Lafay index	2008–2011	Czech Republic
Ishchukova and Smutka (2013)	Agricultural products: cereals such as wheat, barley, and their by-products	RCA (Balassa's index) RTA (Vollrath's index) Lafay index	1998–2010	Russia
Sarker and Ratnasena (2014)	Agri-food sector: wheat, beef, and pork	BRCA, RCA, NRCA	1961–2011	Canada
Narayan and Bhattacharya (2019)	Agricultural exports: wheat, sugar, cotton, and rice	RCA, REC	1981–2012	India

(continued)

Table 2 (continued)

Author(s)	Products	Indicator	Period	Region
Cele et al. (2022)	Dairy industry: butter, powders, liquid milk	NRCA	2015	Ireland
Xia and Dewi (2022)	Coconut oil, virgin coconut oil	Trade specialisation index (TSI) Revealed comparative advantage (RCA)	2010–2020	Indonesia
Sharma et al. (2022)	Fruits and vegetables	RCA, RSCA, NRCA	2010–2020	India

Source: Own composition

6.2 *Regional Studies*

Bojnec and Fertő (2018) intended to explore the comparative advantages of agri-food exports in the European Union. The results indicated that, although the NRCA index was higher than zero for most agro-food items, a substantial percentage of them have a shorter duration, lasting only a reduced number of years. Mizik et al. (2020) aimed to analyse comparative advantage patterns of agriculture in the Commonwealth of Independent States (CIS) based on the RCA index, showing that Moldova, Kyrgyzstan, and Armenia have the highest Balassa indices in the region, and Belarus, Ukraine, and Azerbaijan also have some comparative advantage. The authors also inferred that RCA stability and duration are limited, suggesting a constant need to adapt and assess updated data to bring novelty to the literature. Last but not least, Hoang (2020) identified the agricultural competitiveness of ASEAN countries in global markets, with analyses of the dynamics of these indicators for the period 1997–2015, testing the consistency between trade indices. Finally, country- and sector-specific research was more popular in the agricultural sector, examining developed countries (Ireland, Canada) and Asian developing countries (Iran, India, and Indonesia) as well as Russia.

6.3 *Country-Specific and Sector-Specific Research*

The competitiveness of the Hungarian agriculture and food processing sector was analysed by Fertő and Hubbard (2003) based on RCA for the period 1992–1998. Hungary has been shown to have a comparative advantage in a variety of agri-food products, including animals and meat. Further along, Bakhshinejad and Hassanzadeh (2012) applied the RCA for agricultural products (walnut, almond hazelnut, apple, and orange) in Iran. They suggest that Iran did not have a comparative advantage in exporting these products. Burianová and Belová (2012) investigated the foreign agricultural trade of the Czech Republic in the period 2008–2011. They included the RCA and the Lafay index for their research. They confirmed that the three countries, Germany, Slovakia, and Poland, are the largest trading partners of the Czech Republic. Regarding comparative advantage, Ishchukova and Smutka (2013) explored Russian foreign trade in agricultural products and food for the period 1998–2010. Several measures of comparative advantage were applied, such as the Balassa and Vollrath indices, along with the Lafay index. The Balassa index explored a group of products which has a relatively stable comparative advantage (cereals such as wheat, barley, and their by-products such as wheat bran), products from their processing (cereal preparations), and other products (oilseeds, vegetable oils, and chocolate). The Lafay index showed that Russia has comparative advantages in relation to the CIS and Asian countries due to its geographical position and trade relations. In addition, Sarker and Ratnasena (2014) investigated the international competitiveness of the wheat, beef, and pork sectors in Canada and determined the drivers of competitiveness. They pointed out that Canada has competitiveness in the wheat sector, but it is not true for the beef or pork sectors. They concluded that agricultural policies in Canada did not deliver the projected benefits, and the NRCA index did not separate the contributions of market-related and non-market failures to the competitive performance of agriculture in Canada. Among others, Narayan and Bhattacharya (2019) analysed the relative competitiveness in India covering exports of sugar, rice, wheat, and cotton, each product relative to other exports. The findings show that India's Green Revolution improved relative export competitiveness in rice and cotton and encouraged agricultural investment. They emphasised that trade agreements are also important determinants of relative export competitiveness, as the World Trade Organization (WTO) had a positive effect on relative export competitiveness in rice. The revealed comparative advantage (RCA), symmetric comparative advantage (RSCA), and normalised RCA (NRCA) indices were calculated by Sharma et al. (2022), in order to measure the indices of selected fruits and vegetables for the period 2010–2020 in India and its competing countries. The results revealed an improvement in India's comparative advantage during the period and indicated that India faces competition from countries such as China, Indonesia, Brazil, Thailand, Argentina, and the European Union for vegetables and certain roots and tubers. In addition, countries such as Mexico, Indonesia, Brazil, and Thailand have been identified as the main competitors of India in fruit and nuts, and the peel of citrus fruit or melons. Xia and Dewi (2022) analysed the trade and competitiveness of Indonesian coconut oil in the international market employing the trade specialisation index (TSI) and revealed comparative advantage (RCA), which is used to analyse the competitiveness of Indonesian coconut oil. The authors concluded that the trade specialisation of Indonesian coconut oil is in the development stage, which means that Indonesia is likely to be a coconut oil exporting country in the future. Turning back to Europe, Cele et al. (2022) examined the competitiveness of the Irish dairy sector relative to member states of the European Union. Although Irish dairy products, such as butter and powders, have shown growth potential in competitiveness after milk quota, other products like cheese and liquid milk have declined in competitiveness according to the key export competitiveness indicators used. Regarding the indices analysed in the selected literature, among others, RCA (15), NRCA (5), and RTA (4) were the most popular techniques, while more advanced indices such as the RSCA (2), RC (1), and Lafay (2) indices were the least applied by the literature (Fig. 2).

Most studies on competitiveness analysed the agricultural sector principally in developed regions of the world such as Europe, Canada, or the United States. Other authors focused mainly on Asian countries (Iran, India, Indonesia) in developing regions. Providing a detailed literature review on the competitiveness of agri-food trade, Mizik (2021) confirmed that although the European Union is overrepresented in the literature, new EU candidates and other important trading partners of the EU, such as Canada, the ASEAN countries, or China, were narrowly investigated (Fig. 3).

However, many studies address competitiveness in developing regions, such as Latin America and the Caribbean, and such analysis is scarce compared to other regions in the world. This was the most important research gap identified in this section.

7 Competitiveness and Comparative Advantages in Latin America and the Caribbean

This section reviews the limited literature on LAC trade competitiveness and its main empirical findings provided by empirical studies. Only a few relevant published materials (6) can be identified that address comparative advantage analysis in

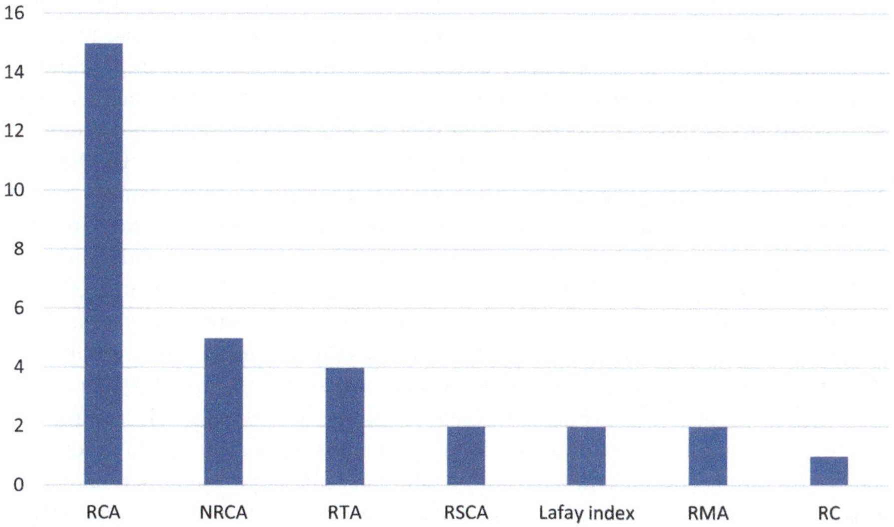

Fig. 2 The popularity of RCA-based indices applied by the literature. Source: Own composition

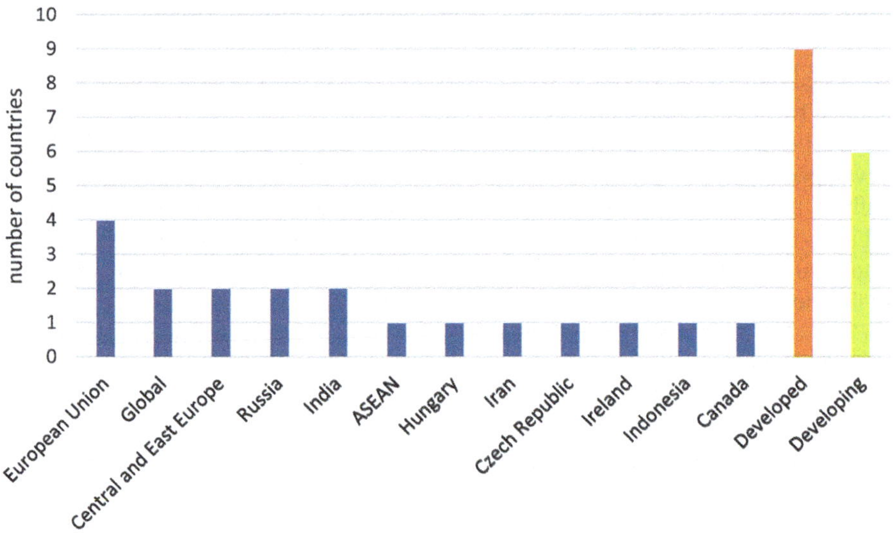

Fig. 3 Frequency of countries and regions analysed by the literature. Source: Own composition

Latin America and the Caribbean region with the help of the Google Scholar search engine (Table 3).

While Beyene (2015) investigated trade between BRICS (Brazil, Russia, India, China, and South Africa) and Sub-Saharan Africa, Peguero et al. (2019) researched Central America and the Caribbean. Borges Aguiar and Balogh (2022) discovered all the agricultural trade patterns and export competitiveness. Finally, Galle et al. (2020) and Zdráhal et al. (2021) concentrated on the topic of the revealed comparative advantages of the agri-food sector in Brazil. To summarise the literature, Beyene (2015) examined economic integration through trade between BRICS countries and Sub-Saharan Africa via comparative advantages with respect to the exportation of merchandise, including food and agricultural raw materials. He concluded that Sub-Saharan Africa is more competitive than BRICS in the exportation of agricultural raw materials and food and economic integration between BRICS and Sub-Saharan Africa favourably influences peace and stability in the regions. Furthermore, Jámbor and Babu (2016) concluded that Latin American countries generally excel in meat; thus, such products and the food security of the LAC region are expected to improve in the long run. Agricultural production in Central America and the Caribbean was researched by Peguero et al. (2019). They suggested that foreign trade has been fundamental to the growth of the agricultural sector in Central America and the Caribbean. The region has a positive trade balance in terms of agricultural products. This region surpasses the export of tropical products such as vegetables, fruits, coffee, sugar, aquaculture products, and tobacco, but has a deficit in the production of cereals, oil crops, animal products, cotton, and fats. The largest commercial partner of the region was the United States, which benefits from diversified year-round tropical fruit and vegetable imports. In recent decades, the export of tropical fruits

Table 3 Studies on comparative advantages in Latin America and the Caribbean agri-food trade

Author	Products	Indicator	Period	Findings
Beyene (2015)	BRICS and Sub-Saharan Africa (agricultural raw materials, foods, and other industrial products)	RCA	1995–2010	Sub-Saharan Africa is more competitive than BRICS in the exportation of agricultural raw materials and food
Jámbor and Babu (2016)	Global agriculture including LAC	RCA	1991–2014	Latin American countries generally excel in meat products. Food security in the Latin America and Caribbean region is expected to improve in the long run
Galle et al. (2020)	Brazilian poultry trade	RCA, SRCA	2009–2016	Comparative advantage during the whole period, the sector is competitive, Brazil has lost its advantages and has been decreasing its indexes
Zdráhal et al. (2021)	Brazilian agri-food sector (cereal meal, flours, live animals)	RCA, Lafay index (LFI), NRCA, stability of the distribution	1995–2017	Other cereal meal and flours maintain weak, while live animal products gain weak comparative advantages; the external shape of agri-food specialisation has strengthened since the early 2000s; the trade shifted more towards China
Borges Aguiar and Balogh (2022)	LAC agri-food sector (different products)	RCA, SRCA, RTA, RC	1995–2019	Brazil, Argentina, and Mexico were the leading exporters; the highest RCA, SRCA, and RTA were found in Guatemala; RC was found in Argentina. Fruit and nuts and the peel of citrus fruits or melons were the most competitive in the worldwide market
Peguero et al. (2019)	Central America and the Caribbean Agricultural Production (tropical products)	RCA	1991–2016	Foreign trade has been fundamental to the growth of the CAC agricultural sector. The CAC excels in the export of tropical products (vegetables, fruits, coffee, sugar, aquaculture products, tobacco) but has a deficit in the production of cereals, oil crops, animal products, cotton, and fats

Source: Own composition

and vegetables from the region increased not only in volume and value but also in diversity. They suggested that since tropical products exported to the United States increased by 12% between 2000 and 2018, the region should take advantage of its privileged tropical position to increase the variety of fruits and vegetables produced to reduce dependence on a limited array of commodities. The competitiveness of poultry meat exports in Brazil and Poland was tested by Galle et al. (2020), using the indices of revealed comparative advantage and the revealed symmetric comparative advantages, exported by the main producers in the period from 2009 to 2016. The results indicated a comparative advantage in Brazilian exports, in the entire period. In relation to the United States and China, there was an advantage in 2 years, but at distinct moments, while in the others there was a comparative disadvantage. Brazil has lost its representativeness and has been decreasing its indices, as those of other countries. Zdráhal et al. (2021) investigated the Brazilian revealed comparative advantages of agri-food products for the period 1995–2017. The results support the argument that the Brazilian agri-food trade was formed by comparative advantages of specific agri-food sectors. Since 2000 and when the trade shifted more towards China, the external shape of agri-food specialisation strengthened. The trade pattern was stable according to the revealed comparative (dis)advantage of particular products, and more changes occurred in each product's score and ranking of products. Recently, Borges Aguiar and Balogh (2022) explored LAC agricultural trade patterns and export competitiveness through the analysis of the revealed comparative advantage index and its modifications (SRCA, RTA, and RC) in the agricultural sector between 1995 and 2019. By examining the characteristics of LAC's agri-food trade, they stated that Brazil, Argentina, and Mexico were the leading exporters between 1995 and 2019. Calculating the Balassa indices revealed that among the main agricultural exporters in LAC, Guatemala, Uruguay, and Ecuador had the highest comparative advantages (RCA) in all periods analysed by Borges Aguiar and Balogh (2022). They show that the highest RCA, SRCA, and RTA were found in Guatemala, while the highest RC was calculated for Argentina. Fruit and nuts, citrus fruit, or melons produced in LAC were the most competitive in the worldwide market. In terms of agriculture, Brazil was the country with the highest research interest in the LAC countries. Regarding agricultural product groups, fruits, cereals, and animal products were the most surveyed. The literature provided mixed results on which LAC agricultural products or certain countries of LAC have comparative advantages in the agro-food trade (Table 4). The literature suggests that LAC has a comparative advantage in fruit, especially tropical fruits and nuts, citrus, melons, vegetables, coffee, sugar, aquaculture, tobacco, and some live animals (e.g. poultry) on the global market. In contrast, many scholars believed that other regions

Table 4 Results for comparative advantage of agri-food products in LAC provided by the literature

Country	Comparative advantage	Author
Brazil (BRICS)	Agricultural raw materials and food products	Beyene (2015)
Brazil	Poultry products	Galle et al. (2020)
LAC (Guatemala, Argentina)	Fruit and nuts, peel of citrus fruit, or melons	Borges Aguiar and Balogh (2022)
Central America and the Caribbean	Tropical products (e.g. vegetables, fruits, coffee, sugar, aquaculture products, tobacco)	Peguero et al. (2019)
LAC	Meat products	Jámbor and Babu (2016)
Brazil	Cereal meal, flours, live animals	Zdráhal et al. (2021)

Source: Own composition

of the world were more competitive than the LAC region in agricultural and food products (Beyene, 2015; Jámbor & Babu, 2016).

In general, some disadvantages weaken the agricultural competitiveness of LAC. Small and medium-sized enterprises (SMEs) in the agri-food sector face limited access to credit and financing, restricting their ability to invest in technology, productivity, and competitiveness. Many LAC countries have moderate investments in R&D which results in low innovation and competitiveness in the agri-food sector. The region has a moderate capacity to implement effective sanitary and phytosanitary measures, which can slow market access for agri-food products. In addition, the LAC region faces trade barriers, such as high tariffs and non-tariff measures (sanitary and phytosanitary restrictions), which reduce its competitiveness in the global market. Finally, the LAC region has underdeveloped infrastructure systems that complicate the transportation of agricultural products to markets efficiently and at a reasonable cost. To increase its competitiveness in the global agri-food market, the LAC region needs to address these challenges through investment in transportation and logistics systems, financing and access to credit for SMEs, research and development, sanitary and phytosanitary measures, and trade facilitation. In addition, the region can take advantage of its comparative advantages, such as its favourable climate, diverse natural resources, and large land area, to increase its competitiveness and become a major player in the global agri-food market. Over the last 15 years, the trade performance of Latin America and the Caribbean has shown less dynamism and greater volatility. In recent years, aside from a few exceptions, LAC economies have become less competitive externally, particularly in the intraregional market. Therefore, it is essential to provide a new stimulus in international integration strategies, focusing on regional integration (Inter-American Development Bank, 2022). Recurring global shocks that have impacted the region's trade performance are signs of a medium-term trend towards instability. The Russian-Ukrainian conflict caused a new shock because Russia is the world's largest gas exporter, the second largest oil exporter, and the major global fertiliser supplier, while both Russia and Ukraine are significant producers of wheat, sunflower, and maize (Inter-American Development Bank, 2022). In the first half of 2022, the overall commodity price index of LAC grew 49.9%, with energy up by 99.3% and fertilisers by 109.5%. Meanwhile, non-energy products grew at a substantially lower rate (14.9%), and the increase for food and beverages (22.9%) was markedly higher than that for agricultural commodities (6.8%) and metals (2.3%). In the short term, the outlook for trade in LAC will depend primarily on the dynamics of commodity prices (Giordano et al., 2022).

8 Cooperation as a Solution for Agricultural Competitiveness

Trade cooperation can be identified as one of the potential solutions to increase trade competitiveness and achieve sustainable development in the agricultural sector. Free trade agreements can be an important tool for agricultural development in LAC countries. In terms of agricultural trade, MERCOSUR (Bienvenidos al Mercado Común del Sur), also called the Southern Common Market, which includes Argentina, Brazil, Paraguay, and Uruguay, has a key role in promoting free trade and trade development within LAC countries. In 2022, the MERCOSUR group comprised 295 million people with a combined GDP of nearly USD2 trillion (Castillejo et al., 2022). Behind MERCOSUR, there is another cooperation in the LAC region, such as the Andean Community (which includes Bolivia, Colombia, Ecuador and Peru), a free trade area to create a customs union comprising South American countries. The Latin American Integration Association (ALADI) includes 12 countries that seek economic cooperation among its members. In 2001, ALADI signed an agreement with the Andean Community that aimed to facilitate further integration of the Mercosur. The Caribbean Community (CARICOM) includes 15 countries in the Caribbean region with the objective of economic integration, foreign policy coordination, human and social development, and security. Another cooperation the Central American Integration System or SICA (members are Belize, Costa Rica, El Salvador, Guatemala, Honduras, Nicaragua, Panama, and the Dominican Republic) has the objective of bringing Central America and the Dominican Republic as a region of peace, freedom, democracy, and development. In conclusion, free trade initiatives and trade cooperation play a crucial role in promoting economic growth and development in the LAC region by reducing trade costs, improving access to global markets, and increasing trade flows between trading partners. It can help countries overcome trade barriers and promote more efficient and effective trade relations.

9 Summary

The LAC region has great potential to significantly increase the competitiveness of the agri-food trade and become a leader in the global market. However, this region faces several challenges that hinder its competitiveness. Despite the global relevance of the topic, research on the patterns and dynamics of competitiveness in Latin America and the Caribbean is scarce compared to other regions of the world. At the macro level, there is a lack of a commonly accepted definition of competitiveness and comparative advantages. The definition of trade competitiveness provided by Jámbor and Babu (2016) is associated with the concept of comparative advantage, which means the country's ability to produce goods or services at a lower opportunity cost than its trading partners. Competitive performance can be determined by different measurements based on indicators such as revealed comparative advantage, market performance, trade success, and others. The revealed comparative advantage (RCA) indices formulated by Balassa (1965) and their modified versions (RMA, RTA, RC, RSCA, NRCA, and Lafay index) are the most widely used indicators for trade-induced competitiveness of nations. Although RCA indexes are usually not consistent with theoretical concepts of comparative advantage, they are appropriate to evaluate the differential effect of changes in trade barriers between producers of different products. In the past few decades, a large number of studies (Mizik, 2021) have been published on analysing comparative advantages in different agricultural sectors. The first group of studies available in the economics literature analysed the competitiveness of agricultural trade at the global level followed by regional and country-level studies. Regarding the indices analysed, among others, RCA, RTA, and NRCA were the most popular techniques, while more advanced indices such as the Lafay index or RC were less frequently applied by the literature. Most studies on competitiveness aimed to analyse the agricultural sector principally in developed regions of the world such as Europe, Canada, or the United States. Other authors focused mainly on Asian countries (Iran, India, and Indonesia) in developing regions. Providing a detailed review of the literature on agri-food trade competitiveness, Mizik (2021) confirmed that although the European Union is overrepresented in the literature of comparative advantages, new EU candidates and other important trading partners of the EU, such as Canada, China, or the ASEAN countries, were also widely investigated. By contrast, the number of studies addressing competitiveness in developing regions, such as Latin America and the Caribbean, is scarce compared to other regions in the world. Only six relevant articles were identified that analysed comparative advantage in Latin America and the Caribbean region with the help of the Google Scholar search engine. While Beyene (2015) investigated trade between BRICS and Sub-Saharan Africa, Peguero et al. (2019) researched Central America and the Caribbean. Borges Aguiar and Balogh (2022) discovered trade patterns and export competitiveness in agriculture, and finally, Galle et al. (2020) and Zdráhal et al. (2021) explored revealed comparative advantages of the agri-food sector in Brazil. Regarding agricultural product groups, fruits, cereals, and animal products were surveyed the most in LAC. The literature provided mixed results on which LAC agricultural products or certain countries of LAC have comparative advantages in the agro-food trade. The literature suggests that LAC has a comparative advantage in fruit, especially tropical fruits and nuts, citrus, melons, vegetables, coffee, sugar, aquaculture, tobacco, and some live animals (e.g. poultry) on the global market. On the contrary, many scholars supposed that other regions of the world were more competitive than the LAC region in agricultural and food products (Beyene, 2015; Jámbor & Babu, 2016). In general, there are some disadvantages that weaken the agricultural competitiveness of LAC. Many countries in the LAC region generally have small investments in R&D which result in low innovation and competitiveness in the agri-food sector. The LAC region has underdeveloped infrastructure systems, making it difficult to transport agricultural products to markets efficiently and at a lower cost. The LAC region needs to address these challenges through investment in transportation and logistics systems, financing and access to credit for SMEs, research and development, and trade facilitation, to increase its competitiveness in the global agri-food market. Over the last 10 years, exports from Latin America and the Caribbean have been less dynamic and more volatile. The recurring global shocks (COVID-19, Russian-Ukrainian conflict) that have affected regional trade performance point to a trend of instability in the medium term. Trade cooperation in LAC such as MERCOSUR, ALADI, CARICOM, and SICA can be the potential solutions to increase trade competitiveness and achieve sustainable development in the agricultural sector. Finally, trade initiatives play a crucial role in promoting economic growth and development in the LAC region by reducing trade costs, improving access to global markets, and increasing the export of agricultural products.

References

Bakhshinejad, M., & Hassanzadeh, A. (2012). Comparative advantage of selected agriculture products in Iran: A revealed comparative advantage assessment. *Agricultura tropica et subtropica, 45*(1), 28–31. https://doi.org/10.2478/v10295-012-0004-9

Balassa, B. (1965). Trade liberalisation and "revealed" comparative advantage. *The Manchester School, 33*(2), 99–123. https://doi.org/10.1111/j.1467-9957.1965.tb00050.x

Beyene, H. G. (2015). BRICS and sub-Saharan Africa trade interdependence and peace: Evidence based on revealed comparative advantages. *Journal of Law, Society and Development*, 2(1). Accessed February 10, 2023, from https://hdl.handle.net/10520/EJC186388

Bojnec, S. (2001). Trade and revealed comparative advantage measures: Regional and Central and East European Agricultural Trade. *Eastern European Economics, 39*(2), 72–98. https://doi.org/10.1080/00128775.2001.11040990

Bojnec, S., & Fertő, I. (2017). The duration of global agri-food export competitiveness. *British Food Journal, 119*(6), 1378–1393. https://doi.org/10.1108/BFJ-07-2016-0302

Bojnec, S., & Fertő, I. (2018). Drivers of the duration of comparative advantage in the European Union's agri-food exports. *Agricultural Economics (Zemědělská ekonomika), 64*(2), 51–60). Czech Academy of Agricultural Sciences. https://doi.org/10.17221/173/2016-agricecon

Bojnec, Š., & Fertő, I. (2019). European Union countries agri-food trade structures and main competitors on the internal and global agri-food markets. *Ekonomika Poljoprivrede, 66*(2), 635–650. https://doi.org/10.5937/ekopolj1902635b

Borges Aguiar, G. M., & Balogh, J. M. (2022). Analysis of the Competitiveness in the Agri-food sector: The case of Latin America and the Caribbean Region. *Competitio, 21*(1–2), 92–117. University of Debrecen. https://doi.org/10.21845/comp/2022/1-2/2

Burianová, J., & Belová, A. (2012). The competitiveness of agricultural foreign trade commodities of the CR assessed by way of the Lafay Index. *Agris on-line Papers in Economics and Informatics, IV*(4), 27–36. Accessed February 14, 2023, from https://www.researchgate.net/publication/282053863_The_competitiveness_of_agricultural_foreign_trade_commodities_of_the_CR_assessed_by_way_of_the_Lafay_Index

Castillejo, C., Fleischner, N., Harrison, Ch., Horwitz, L., Keller, M., & Orbach, J. (2022, August 23). Explainer: What is Mercosur? AS/COA. Accessed February 15, 2023, from https://www.as-coa.org/articles/explainer-what-mercosur

Cele, L. P., Hennessy, T., & Thorne, F. (2022). Evaluating farm and export competitiveness of the Irish dairy industry: Post-quota analysis. *Competitiveness Review: An International Business Journal, 32*(7), 1–20. https://doi.org/10.1108/cr-11-2020-0136

Dalum, B., Laursen, K., & Villumsen, G. (1998). Structural change in OECD export specialisation patterns: de-specialisation and 'stickiness'. *International Review of Applied Economics, 12*(3), 423–443. https://doi.org/10.1080/02692179800000017

De Benedictis, L., Tamberi, M., De Benedictis, L., & Tamberi, M. (2004). Overall specialization empirics: Techniques and applications. *Open Economies Review, 15*(4), 323–346. https://econpapers.repec.org/RePEc:kap:openec:v:15:y:2004:i:4:p:323-346

Ferrarini, B., & Scaramozzino, P. (2011). Indicators and patterns of specialization in international trade. Swiss National Center of Competence Research Working Paper No 2011/10| March 2011. Accessed February 7, 2023, from https://www.wti.org/media/filer_public/3c/45/3c450e2f-21a9-4ce0-846b-f33acb38b7b5/indicators_and_patterns_of_specialization.pdf

Fertő, I. (2008). The evolution of agri-food trade patterns in Central European countries. *Post-Communist Economies, 20*(1), 1–10. https://doi.org/10.1080/14631370701865680

Fertő, I., & Hubbard, L. J. (2003). Revealed comparative advantage and competitiveness in Hungarian agri-food sectors. *The World Economy, 26*(2), 247–259. https://doi.org/10.1111/1467-9701.00520

French, S. (2017). Revealed comparative advantage: What is it good for? *Journal of International Economics, 106*, 83–103. https://doi.org/10.1016/j.jinteco.2017.02.002

Galle, V., Rachor, E., Arruda Coronel, D., Machado Pinto, N. G., & Costa, N. L. (2020). Vantagem comparativa revelada da indústria da carne de frango brasileira e dos principais players (2009-2016). *Revista Eletrônica Científica, Da UERGS, 6*(1), 42–53. https://doi.org/10.21674/2448-0479.61.42-53

Giordano, P., Campos, R., & Michalczewsky, K. (2022, November). Trade and Integration Monitor 2022: Shockwaves: Latin America and the Caribbean Facing Global Trade Turmoil. Trade and Integration monitor. Inter-American Development Bank. https://doi.org/10.18235/0004540.

Hoang, V. V. (2020). Investigating the agricultural competitiveness of ASEAN countries. *Journal of Economic Studies, 47*(2), 307–332. https://doi.org/10.1108/JES-10-2018-0366

Hoen, A. R., & Oosterhaven, J. (2006). On the measurement of comparative advantage. *The Annals of Regional Science, 40*(3), 677–691. https://doi.org/10.1007/s00168-006-0076-4

Inter-American Development Bank. (2022, November 10). *Exports from Latin America and the Caribbean grow more than world trade despite slowdown.* Accessed February 10, 2023, from https://www.iadb.org/en/news/exports-latin-america-and-caribbean-grow-more-world-trade-despite-slowdown

Ishchukova, N., & Smutka, L. (2013). Revealed comparative advantage of Russian agricultural exports. *Acta Universitatis Agriculturae et Silviculturae Mendelianae Brunensis, 61*(4), 941–952. https://doi.org/10.11118/actaun201361040941

Jámbor, A., & Babu, S. (2016). *The competitiveness of global agriculture: Policy lessons for food security*. Springer. https://doi.org/10.1007/978-3-319-44876-3_6

Jámbor, A., Tóth, J., & Köröshegyi, D. (2018). Competitiveness in the trade of spices: A global evidence. *Bulgarian Journal of Agricultural Science, 24*(5), 729–736. http://agrojournal.org/24/05-01.pdf

Lafay, G. (1992). The measurement of revealed comparative advantages. In M. G. Dagenais & P. A. Muet (Eds.), *International trade modeling*. Chapman and Hill.

Latruffe, L. (2010). *Competitiveness, productivity and efficiency in the agricultural and agri-food sectors, OECD Food, Agriculture and Fisheries Papers, No. 30*. OECD Publishing. https://doi.org/10.1787/5km91nkdt6d6-en

Liesner, H. H. (1958). The European common market and British industry. *The Economic Journal, 68*(270), 302–316. https://doi.org/10.2307/2227597

Mizik, T. (2021). Agri-food trade competitiveness: A review of the literature. *Sustainability, 13*, 11235. https://doi.org/10.3390/su132011235

Mizik, T., Gál, P., & Török, Á. (2020). Does agricultural trade competitiveness matter? The case of the CIS countries. *Agris On-Line Papers in Economics and Informatics, 12*(1), 61–72. https://doi.org/10.7160/aol.2020.120106

Narayan, S., & Bhattacharya, P. (2019). Relative export competitiveness of agricultural commodities and its determinants: Some evidence from India. *World Development, 117*, 29–47. https://doi.org/10.1016/j.worlddev.2018.12.013

Peguero, F., Zapata, S., & Sandoval, L. (2019). Agricultural production of Central America and the Caribbean. *Choices, 3rd Quarter 2019, 34*(3), 1–10. https://www.jstor.org/stable/10.2307/26964937

Reyes, G. E., & Useche, A. J. (2019). Competitiveness, economic growth and human development in Latin American and Caribbean countries 2006-2015. *Competitiveness Review: An International Business Journal, 29*(2), 139–159. https://doi.org/10.1108/cr-11-2017-0085

Sarker, R., & Ratnasena, S. (2014). Revealed comparative advantage and half-a-century competitiveness of Canadian agriculture: A case study of wheat, beef, and pork sectors. *Canadian Journal of Agricultural Economics, 62*(4), 519–544. https://doi.org/10.1111/cjag.1205

Sharma, A., Kathuria, L. M., & Kaur, T. (2022). Analyzing relative export competitiveness of Indian agricultural food products: A study of fresh and processed fruits and vegetables. *Competitiveness Review: An International Business Journal.*, Ahead-of-print. https://doi.org/10.1108/CR-03-2022-0039

Statista. (2023a). *Smart agriculture in Latin America – statistics & facts*. Accessed February 8, 2023, from https://www.statista.com/topics/7206/smart-agriculture-in-latin-america/#topicOverview

Statista. (2023b). *Statistics about agriculture in LAC. Share of value added by the agricultural sector to the gross domestic product (GDP) in Latin America and the Caribbean in 2020, by country*. Accessed February 9, 2023, from https://www.statista.com/statistics/1079120/latin-america-caribbean-agriculture-share-gdp/

Statista. (2023c). *Statistics about agriculture in LAC. Employment in the agricultural sector as share of total employment in Latin America and the Caribbean in 2020, by country or territory*. Accessed February 9, 2023, from https://www.statista.com/statistics/1082252/latin-america-caribbean-share-employment-agriculture-country/

Török, A., & Jámbor, A. (2016). Determinants of the revealed comparative advantages: The case of the European ham trade. *Agricultural Economics – Czech, 62*(10), 471–481. https://doi.org/10.17221/177/2015-AGRICECON

Vollrath, T. L. (1991). A theoretical evaluation of alternative trade intensity measures of revealed comparative advantage. *Weltwirtschaftliches Archiv, 127*(2), 265–280. https://doi.org/10.1007/BF02707986

World Bank. (2023a). *World Integrated Trade Solution (WITS). LCN 10 trade statistics: Exports, imports, products, tariffs, GDP and related development indicator*. Accessed February 10, 2023, from https://wits.worldbank.org/CountryProfile/en/LCN 10

World Bank. (2023b). *World Integrated Trade Solution (WITS). Latin America & Caribbean Food Products Exports by country and region in US$ thousand 2020*. Accessed February 10, 2023, from https://wits.worldbank.org/CountryProfile/en/Country/LCN/Year/2020/TradeFlow/Export/Partner/all/Product/16-24_FoodProd# 2022

Xia, L., & Dewi, S. (2022). Analysis of trade specialization and competitiveness of Indonesian coconut oil in the international market (2010-2020). *Open Journal of Business and Management, 10*, 245–262. https://doi.org/10.4236/ojbm.2022.101015

Yu, R., Cai, J., & Leung, P. S. (2009). The normalized revealed comparative advantage index. *Annals of Regional Science, 43*, 267–282. https://doi.org/10.1007/s00168-008-0213-3

Zdráhal, I., Hrabálek, M., Kadlec, P., & Krpec, O. (2021). Brazil's comparative advantages and specialization dynamics in agri-food trade. *AGRIS On-line Papers in Economics and Informatics, 13*(2), 121–139. https://doi.org/10.7160/aol.2021.130210

Member Preferences for Cooperative Attributes: A Best-Worst Scale Analysis

Tilemachos Manousakis, Panagiota Sergaki, Nikolaos Kalogeras, and Gert van Dijk

1 Introduction

Cooperatives are considered to be two-layered enterprises due to the "dual objective" of their economic dimension. That is, "the goal of a cooperative is achieved when it successfully generates wealth for both the cooperative and its members" (Van Dijk et al., 2019, p. 25). The viability of a cooperative enterprise due to its collective nature is highly dictated by the level of commitment of its members (Fulton & Giannakas, 2001). However, the globalization of markets intensified the competition between member-invested enterprises and other IOFs (Novkovic, 2008). Other than the problems related to their economic performance, cooperatives often face problems related to their social dimension (Grashuis, 2018).

Evidently, what is crucial for addressing the problems related to the social dimension of the cooperative is the construction of strong social capital. In a cooperative organizational framework, social capital constitutes both a resource and a foundational principle for the enterprise (Liang et al., 2015). Additionally, Valentinov

T. Manousakis (✉)
Business Economics and Management, Mediterranean Agronomic Institute of Chania, Chania, Greece

P. Sergaki
Aristotle University of Thessaloniki, Thessaloniki, Greece

N. Kalogeras
Sustainable International Business (SIB) Research Center, Domain of International Business & Communication, Zuyd University, Heerlen, The Netherlands

Commodity Risk Management Expertise Center (CORMEC), Marketing & Consumer Behaviour Group, Wageningen University & Research (WUR), Wageningen, The Netherlands

G. van Dijk
Nyenrode Business University, Amsterdam, The Netherlands

K. Mattas et al. (eds.), *Value Chain Dynamics in a Biodiverse Environment*, Cooperative Management, https://doi.org/10.1007/978-3-031-49845-9_3

(2004) highlights the importance of social capital, especially for cooperative governance structures. The role of social capital for a cooperative structure is considered to be equivalent in terms of importance as the financial and physical capital for an IOF.

The level of commitment of different members is subject to their diverse preferences for several cooperative functions. In that sense, members may have different motives for committing or not to the collective enterprise (Kalogeras et al., 2007, 2009). Knowledge on the preferences of members regarding the cooperative business may result in enhanced commitment. Inspired by Kalogeras et al. (2007, 2009), who first studied heterogeneous preferences of cooperative members in the Netherlands, and considering the challenges encountered in previous work on attributes affecting the members' commitment (Manousakis et al., 2021), we further examine the relative importance of preferences for an expanded list of cooperative attributes that drive the commitment of cooperative members using an advanced quantitative approach. Thus, the main objective of this chapter is to examine the preferences of cooperative members for several cooperative attributes and the diversity in their preference structure.

To address this objective, we firstly evaluate the perceived relative importance that members attach to an expanded list of cooperative attributes by applying a best-worst scale (BWS) analytical approach. Secondly, we examine the diversity in member preferences using a latent class analysis (LCA) framework.

The contribution of this study is demonstrated by the use of an advanced quantitative method on the revealed preferences of cooperative members for cooperative attributes, a topic which has not received much explicit attention in the academic literature on cooperatives using the decision context of Greece. Using data provided by members of a prominent agricultural cooperative in Greece, with more than a hundred years of commercial history, and relating them to the theoretical background of international literature, the current quantitative study may provide crucial information on the identification of the members' preferences for cooperative attributes that are affecting their commitment to the collective enterprise.

The chapter is organized as follows. Section 2 provides a literature background of the components that are needed for building the member's active membership, namely (a) cooperative culture, (b) open communication, (c) trust, (d) involvement, and finally (e) willingness to be active. The methodology of this study is described in Sect. 3. Section 4 presents the interpretation of the findings according to the information provided by the BWS and LCA analyses. Section 5 presents the conclusions of this study. Finally, Sect. 6 discusses the limitations of the study along with some recommendations for future research.

2 Theoretical Framework

2.1 Active Membership in Cooperatives

In order to identify important and relevant cooperative attributes, we first establish a theoretical framework regarding the significance of committed members for the prosperity of the cooperative as well as for the members as individual agents (Verhees et al., 2015). According to Abrisham (2011) and Barraud-Didier et al. (2012), the members' active membership may have a significant impact on their sense of commitment to the cooperative. Members' active membership can be conceptualized as a mixture of social attributes that are linked with the individual characteristics of the member as well as their relationship with the cooperative and the other members. Figure 1 depicts the social attributes that assist in building the members' active membership.

Therefore, we are going to present the relationships between the social factors that result in the demonstration of strong active membership and analyze them as five distinct sections, as follows: (1) cooperative culture, (2) open communication, (3) trust, (4) involvement, and (5) willingness to be active. Below, we elaborate on each of these factors.

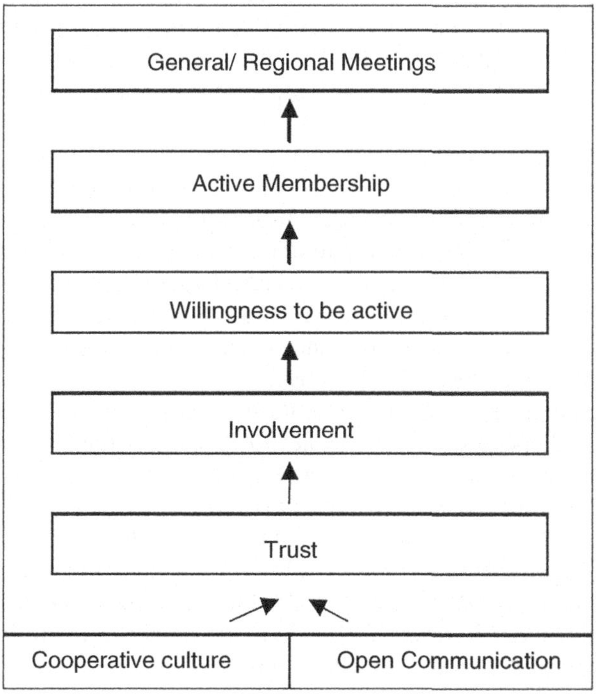

Fig. 1 Social attributes that build up active membership. Retrieved from: Verhees et al. (2015)

2.1.1 Cooperative Culture

Members' cooperative culture is the combination of the values, norms, and beliefs that characterize an individual and steer his/her decisions toward believing in the significant value of the cooperative for his/her survival, as well as the survival of the community he/she belongs to (Verhees et al., 2015). When a cooperative member understands the significance of the cooperative's prosperity and identifies himself/herself as part of the means to achieve it, he/she might become more interested in participating in the formulation of the strategies, show more trust in the Board of Directors (BoD), and finally become more committed to the collective enterprise (Borgen, 2001; Osterberg & Nilsson, 2009).

A member's cooperative culture, and consequently his/her commitment to the collective enterprise, is often defined by his/her personal characteristics. The commitment level of members may depend on the nature of their personal characteristics; for example, the members' commitment level declines as the level of formal education of a member increases (Trechter et al., 2002). In addition, when a cooperative member participates in the BoD and serves in different committee works, as well as receiving training and education on cooperative practices and strategies, the level of his/her commitment to the cooperative increases. Cooperative members who are involved in these activities can become more committed than others (Gray & Kraenzle, 1998; Trechter et al., 2002).

Regarding the members' age, older members seem to be highly loyal to their cooperatives because they have a sense of pride in the ownership of their cooperative (Fulton & Adamowicz, 1993). Staatz (1987) argues that older farmers, who are approaching retirement age, have low levels of commitment compared to younger members. Yet, Hakelius (1996) found that young farmers see their cooperative commitment as a means to obtain economic advantages, whereas older members view it as a way of showing solidarity with peers, while economic benefit is secondary.

Further, the farm size of the members (often measured in terms of annual turnovers or number of employees) has an important influence on the commitment level of members to their cooperative. A farmer with a larger farm size would more likely become more involved in the cooperative (Gray & Kraenzle, 1998). The farm size also seems to be a reason for the presence of heterogeneity in preferences among groups of members (Kalogeras et al., 2009).

Finally, regarding the size of different cooperatives, in terms of numbers of members or membership size, one may acknowledge differentiated levels of member commitment, and this is due to the different individual preferences of cooperative members toward cooperative intra-organizational and strategic attributes (Kalogeras et al., 2009). The coordination of member preferences is essential for the maintenance of cooperative operations and key for the long-term survival and prosperity of the organization. As noted in the literature, increased heterogeneity in members' preferences can result in a decline of member commitment due to conflicting preferences (Fulton & Giannakas, 2001).

2.1.2 Open Communication

Another significant component that assists in inspiring members to become more active is the existence of sufficient and effective communication between the members and the cooperative. Good quality information flow between a cooperative and its members reduces the transaction costs and empowers the member's feelings of belonging to a social group, which is characterized by solidarity relationship bonds (Barraud-Didier et al., 2012). As expressed by Boned and Mutuel (2006), sufficient information provided by a cooperative to its members may constitute a competitive advantage for a collective business, while members may identify market opportunities and may adapt faster to the dynamic changes of the globalized market environment. In addition, good quality communication between the members and the BoD may result in better understanding and cooperation of the members on the achievement of goals and objectives established by the governing instruments of a cooperative (Trechter et al., 2002). Nevertheless, members' involvement through participating actively in governance procedures provides significant individual benefits by reducing the asymmetry of information between the cooperative management and members (Verhees et al., 2015).

2.1.3 Trust

As defined by Verhees et al. (2015), members' belief of trust in their cooperative is crucial in order for members to anticipate the actions made by the cooperative with a "soft heart." In this sense, a member may anticipate a positive outcome from the cooperative business rather than be suspicious and anticipate a negative one. Building strong bonds of trust is highly associated with the achievement of high levels of building social capital as an asset for a cooperative (Chloupkova et al., 2003).

A member's sense of trust in the cooperative constitutes a strong determinant for their demonstrated attitudes, since it is subjected to the level of information asymmetry between the members and their cooperative (Kollock, 2009). Borgen (2001) refers to the asymmetry of information as a strong predictor of the relationship between a cooperative and its members. Members most probably will exhibit more trust in their cooperative when they receive the information relative to collective business functions. This implies that information may be relative to the market prices of products and supplies, and the behavior of the cooperative's clients, among others (Borgen, 2001).

In the context of a social group of individuals sharing common goals and ideas, the success of their cooperation is highly dependent on the nature of the interactions between them and the existence of *reciprocity* (i.e., reciprocal behaviors) (Fowler & Christakis, 2010). Ultimately the existence of "viable" relationships leads to the building of trust among cooperative members, as well as between members and their cooperative. In the field of social psychology, the norm of reciprocity has been utilized by many studies to describe a plethora of phenomena (Avgeris et al., 2018). From a business perspective, reciprocity may be viewed as a factor that aims to enhance social capital within an organization through mutual trust and cooperation (Kontogeorgos et al., 2017). In a cooperative framework, relations developed among members based on reciprocity are found to be fundamental for the empowerment of the collective action norm (Sergaki et al., 2020), and furthermore, on the enhancement of the cooperatives' sustainability, since it can progressively be developed into member commitment and trust (Fowler & Christakis, 2010).

2.1.4 Involvement

The involvement of members in the organizational functions is considered to be of high significance for the vitality of the cooperative in the long term, since involvement assists the transformation of trust in the willingness to be active on behalf of the members (Verhees et al., 2015). Österberg and Nilsson (2009) highlighted the importance of the members' involvement in the general assemblies and member committees of the cooperative, which ultimately enhances the efficiency of the cooperative's governing mechanisms. When members participate actively and are more involved in the cooperative affairs, they are also able to receive information and consequently identify opportunities faster. Sequentially, the identification of opportunities provides a competitive advantage for a cooperative against its rival firms (Boned & Mutuel, 2006).

2.1.5 Willingness to be Active

What is vital for inspiring cooperative members to actively participate in their cooperative affairs is the degree of satisfaction they gain when conducting business with their cooperative. Grashuis and Cook (2018) argue that satisfaction of members positively affects the perceived utility members attain from a diverse set of factors, when they conduct business with their cooperative. The factors that result in increased utility could be the improved price, product quality, input access, well-established marketing channels, and, overall, factors associated with the economic compensation when conducting business with a cooperative. However, members' satisfaction does not only express a member's perceived positive impact in period t, but also predicts the members' behavior in period $t+1$ regarding their commitment to their cooperative. Hernandez-Espallardo et al. (2012) and Arcas-Lario et al. (2014) provided evidence about the positive relationship between member satisfaction and long-term commitment. They note that member satisfaction has a significant impact on the members' willingness to retain their membership in an organization in the long run. Thus, it is vital that the cooperative focuses on understanding the factors that affect their members' satisfaction and acts toward enhancing them.

Resulting from the increased involvement combined with high degrees of satisfaction, the attitude of members that is also crucial for the viability of the cooperative is their active participation in the procedures of governance. On the one hand, members' active participation can possibly lead to higher and better performance of cooperatives (Verhees et al., 2015). For a cooperative to achieve higher percentages of actively participating members in their cooperative's affairs and decision-making procedures, it has to sustain high levels of member satisfaction and member commitment (Dakurah et al., 2005). On the other hand, according to Bijman and Ruben (2005), when members perceive an increase in the gap between the cooperative's vision and their own, their commitment to the collective enterprise is decreased, and consequently their active participation. The decrease of members' commitment in cooperatives may have an impact on the efficiency and continuity of the operations of a collective venture; hence, for member-owned firms, the commitment of their members is vital for their survival (Bijman & Verhees, 2011).

3 Methodology

Accounting for the insights of cooperative experts retrieved by previous scientific work on the exploration of cooperative attributes affecting the member's commitment (Manousakis et al., 2021) and by examining the cooperative literature regarding members' commitment, we identify cooperative attributes that are further used in the BWS (best-worst scale) analysis. The BWS allows for the identification of individual members' perceived relative importance for specific cooperative attributes.

3.1 Participants and Sampling

The questionnaire-based survey was carried out in Zagora, Magnesia, Greece, in the "ZAGORIN" agricultural cooperative during October 2021 to November 2021. The collection of data was carried out through on-site field research in the villages of Zagora, Makryrachi, and Pouri, wherein the lead author spent 30 days. Individual participants were approached in face-to-face meetings in the office facilities of the cooperative. In total, 106 cooperative farmers from the villages of Zagora, Makryrachi, and Pouri participated in the survey. Efforts were made to sustain the sample as representative as possible, subject to the diversity of socio-demographic characteristics, which are presented in Table 1.

The data were collected through face-to-face meetings with individual members. However, the qualitative data gathered from member meetings were not recorded, and they were just utilized intuitively by the author to interpret the quantitative results.

Table 1 Socio-demographic characteristics of the sample

	Total		Zagora		Makryrachi		Pouri	
	N = 106		*N* = 79		*N* = 12		*N* = 15	
	N	%	*N*	%	*N*	%	*N*	%
Gender								
Male	74	69.8%	51	64.6%	12	100%	11	73.3%
Female	32	30.2%	28	35.4%	0	0%	4	26.7%
Age								
Less than 20	1	0.9%	0	0.0%	1	8.3%	0	0.0%
20–29	17	16.0%	11	13.9%	2	16.7%	4	26.7%
30–39	29	27.4%	21	26.6%	3	25.0%	5	33.3%
40–49	18	17.0%	13	16.5%	3	25.0%	2	13.3%
50–65	37	34.9%	30	38.0%	3	25.0%	4	26.7%
More than 65	4	3.8%	4	5.1%	0	0.0%	0	0.0%
Educational level								
High school or below	71	67.0%	49	62.0%	10	83.3%	12	80.0%
Bachelor	30	28.3%	25	31.6%	2	16.7%	3	20.0%
Masters and doctorate	5	4.7%	5	6.3%	0	0.0%	0	0.0%
Monthly income								
Less than 500	7	6.6%	5	6.3%	1	8.3%	1	6.7%
500–1000	50	47.2%	38	48.1%	5	41.7%	7	46.7%
1000–2000	30	28.3%	20	25.3%	3	25.0%	7	46.7%
2000–3000	9	8.5%	8	10.1%	1	8.3%	0	0.0%
More than 30,000	10	9.4%	8	10.1%	2	16.7%	0	0.0%
Household size								
1 person	9	8.5%	7	8.9%	1	8.3%	1	6.7%
2 persons	27	25.5%	20	25.3%	4	33.3%	3	20.0%
3 persons	33	31.1%	24	30.4%	3	25.0%	6	40.0%
4 persons	30	28.3%	26	32.9%	2	16.7%	2	13.3%
5 persons	5	4.7%	2	2.5%	2	16.7%	1	6.7%
More than 5 persons	2	1.9%	0	0.0%	0	0.0%	2	13.3%
Area of land bearing fruits (in m^2)	M=26,603.77		M=27,151.9		M=24,833.33		M=25,133.33	
	(SD=21,450.13)		(SD=23,838.18)		(SD=17,150.85)		(SD=6854.27)	

3.2 Research Method

The BWS method was employed in order to assess the importance rating of each attribute (Lee et al., 2008; Marley & Louviere, 2005). Best-worst scaling, also called the maximum difference scaling method (MaxDiff.), is a quantitative discrete-choice method that measures the subjective value associated with particular objects at an aggregate and/or individual level (Marley & Louviere, 2005). Although the BWS method hasn't been employed in past studies exploring members' preferences in cooperatives, it has become very popular in the food marketing and consumer research field, especially in studies related to the identification of consumers' preferences toward food and beverage attributes (Mueller et al., 2010; Chrysochou et al., 2012b). In this study, we view the cooperative firm as a system of attributes. We assume that members attach utility to these attributes. To the best of our knowledge, this is the first time that BWS analysis has been utilized to identify the cooperative members' perceived importance regarding cooperative attributes.

The reasoning behind our choice of implementing the BWS method for studying member preferences is nested in the advantages that this method provides, particularly the advantages of BWS as explained by Chrysochou et al. (2012a): "…it is free from individual scale usage bias and infers a ratio level importance scale that allows for comparisons across participant segments." In this study, participants were asked to reveal their choice on the attribute that most and least influenced their decision to remain committed to their cooperative.

Furthermore, the sample of the participants was splintered into sub-groups through LCA, which is employed in order to reveal the unobserved heterogeneity of the variables of a given sample (Henry & Muthén, 2010; Chrysochou et al., 2012a; Grymshi et al., 2021). In our study, the participants were splintered into classes relative to their preferences on several cooperative attributes.

3.3 Selected Attributes

Members' active membership can be conceptualized as a mixture of social attributes that are linked with the individual characteristics of members as well as their relationship with the cooperative and the other members (Verhees et al., 2015). In Table 2, the inventory of seven cooperative attributes is presented and the most prominent scholarly works referring to each attribute are mentioned. The formulation of the inventory of cooperative attributes was carried out for the purposes of a previous qualitative study regarding the identification of strategies for enhancing members' commitment in cooperatives (Manousakis et al., 2021). Both studies use an identical research design. The selected attribute resulted from combining cooperative expert interviews and the inferences of the literature review on cooperative attributes that members derive utility from.

At this point, we should acknowledge that the lack of empirical research in the identification of the attributes related to cooperative functions, with a few exceptions such as the studies of Kalogeras et al. (2007, 2009), might have led to a biased choice of attributes, which may not necessarily represent the whole spectrum of the members' motives and preferences for commitment in a collective enterprise. Namely, the cooperative attributes that were utilized in the BWS and LCA analysis were economic benefits, additional services, market access, cooperative governance, communication, control, and cooperative ideology.

Table 2 The inventory of the seven cooperative attributes

Attribute	Explanation	Retrieved from
Economic benefits	Increasing the value of economic benefits for members by providing better prices for their product, cooperative dividends, and improving cost efficiency (better prices for fertilizers, medicines and production materials)	Nishi and Kumar (2011), Liebrand and Ling (2014)
Additional services	Enhancement of additional services provided by the cooperative such as technical support (e.g., hiring experts) and training seminars	Kalogeras et al. (2007), Ishimwe et al. (2016)
Market access	Strengthening the position of cooperatives in the market by increasing the bargaining power with market-oriented strategies (e.g., provide assured marketing channels, build strong brand name, increase the differentiation of products and services)	Van Dijk (1999), Cook & Chaddad (2004), Kalogeras et al. (2007)
Cooperative governance	Reconstruction of the governance structure of cooperatives and cooperative rules (e.g., approaches for how to solve disagreements between members)	Iliopoulos and Theodorakopoulou (2014), Verhees et al. (2015)
Communication	Increase communication between members, as well as between members and the BoD, for reducing information asymmetries (more frequent briefings on market information) and increase economic transparency	Bijman & Verhees (2011), Verhees et al. (2015), Van Dijk et al. (2019)
Control	Increase control and evaluation both from members' committees to the BoD and from the BoD to members, in order to establish a mechanism for ensuring that the cooperative acts on behalf of the members, and vice versa	Trechter et al. (2002), Bhuyan (2007)
Cooperative ideology	Promote the added value of the cooperation and empower the members' sense of cooperative ideology through educational programs	Flecha and Ngai (2014), Morfi et al. (2015), Verhees et al. (2015)

Source: Manousakis et al. (2021)

3.4 Questionnaire Design

The questionnaire assigned to this study was divided into two sections. The first is related to the personal characteristics and attitudes of the participants relative to their household and farm size, as well as to the degree of satisfaction with the cooperative, and the level of production they market through the cooperative. In the second section, the participants were provided with different subsets (see the following section) of the following set of seven cooperative attributes influencing member commitment: "economic benefits," "additional services," "market access," "cooperative governance," "communication," "control," and "cooperative ideology," and were asked to choose the best and the worst attributes in each subset.

The selected cooperative attributes were distributed into seven subsets of three items each. In particular, balanced incomplete block designs were utilized to form

LEAST IMPORTANT	COOPERATIVE ATTRIBUTES	MOST IMPORTANT
	2. Provides additional services	
	4. High quality communication within the Coop	
	6. Good Coop governance model	

Fig. 2 Example of a questionnaire BWS choice set

the best-worst scale questions. Through these experimental designs, which enable the equally frequent presentation of each choice option, the selected attributes subsequently obtain equal co-appearance with each other in the survey design (Louviere et al., 2015). Specifically, the attributes appear in the questionnaire in such a way that all attributes appear equally in the total sets of choices, while each attribute's pair in the same set appears only once. Examples of the BWS choice sets utilized in this study appear in Fig. 2. The participants were asked to choose the most important and the least important attribute for their motivation to remain committed to their cooperative.

LEAST IMPORTANT	COOPERATIVE ATTRIBUTES	MOST IMPORTANT
	2. Provides additional services	
	4. High-quality communication within the coop	
	6. Good coop governance model	

3.5 *Data Analysis*

For the estimation of each attribute's aggregated BWS score, we subtracted the number of times each separate attribute was chosen as "least important" from the number of times the same attribute was chosen as "most important." According to Finn and Louviere (1992), this process is based on the participant's random utility maximization. The differences between the best and the worst values correspond to the utility value of each different attribute, at both the sample and the individual level. The results extracted from the above-mentioned calculations are individual-level BWS scores explicitly for each attribute and are easily comparable with one another across the entire sample. In this study, each attribute appeared a total of three times in the whole questionnaire; therefore, the range of each attribute's individual-level BWS score ranges from −3 to +3.

In order to classify participants into clusters, we utilized the individual-level BWS scores of each attribute in a latent class clustering analysis (LCA) approach. LCA has been utilized extensively in consumer behavior literature (Chrysochou et al., 2012a; Grymshi et al., 2021) for identifying the heterogeneity of consumer preferences toward several product characteristics. In our study, we employ LCA in order to identify the heterogeneity in members' preferences for cooperative attributes.

The Stata/MP.16 software was employed to estimate the LCA models (Grymshi et al., 2021). These scores were used as a ground base for the classification of the resulting classes. Through LCA, subtypes of related cases are identified in a categorical set of observed variables, which are called latent classes (Henry & Muthén, 2010). By reviewing the relative international literature (Vermunt & Magidson, 2002; Chrysochou et al., 2012a; Grymshi et al., 2021), one may realize that latent class clustering analysis has many advantages compared to other traditional cluster analyses (e.g., K-means or hierarchical). Specifically, the advantages LCA yields are as follows: (1) greater evaluation of modeling fitness though its probabilistic approach is allowed, (2) there are a variety of diagnostic criteria and fit statistics for defining the clusters, (3) LCA does not require linearity, normal data distribution, and variance homogeneity, and (4) different scale types, such as categorical, continuous, or any combination of these, can be utilized.

For defining the cluster characteristics of the participants belonging to each subgroup, a chi-square test of association was employed in order to reveal the association of the characteristics of the participants with the above-mentioned members' preference clusters, as provided by LCA. The test was estimated utilizing respondent cooperative members' socio-demographics data, as well as their opinions on their perceived satisfaction with the cooperative organization. The chi-square tests belong to the Karl Pearson family tests and are simple and computable. In addition, they are utilized to examine the independence or association between two categorical variables or the goodness of fit of a given sample to the distribution of a known population (Franke et al., 2012).

4 Results and Discussion

4.1 Perceived Importance of the Cooperative's Attributes

Table 3 shows the aggregated BWS scores across the total sample, the mean of the individual-level BWS scores, and its standard deviation. For the total sample, the most important cooperative attribute is the economic benefits provided by the coop (0.62) followed by the provision of market access (0.32). On the opposite end of the spectrum, communication techniques and the cooperative governance model are the attributes that scored the lowest among the others, respectively (−0.24), followed by cooperative ideology (−0.22), control mechanisms (−0.13), and additional services (−0.12).

Table 4 presents the average individual BWS scores of the cooperative's attributes for the total sample of our study as well as for the distinct regional locations

Table 3 Aggregated and individual-level BWS scores for each cooperative attribute

Attributes	Total best	Total worst	Aggregated BWS	Mean of individual-level BWS	SD of individual-level BWS
Economic benefits	153	87	66	0.62	1.97
Additional services	95	108	−13	−0.12	1.48
Market access	119	85	34	0.32	1.69
Communication	84	109	−25	−0.24	1.56
Cooperative ideology	99	122	−23	−0.22	1.74
Cooperative governance model	89	114	−25	−0.24	1.47
Control mechanisms	103	117	−14	−0.13	1.68

of the cooperative members. Additionally, the ranking of each different functional cooperative attribute is presented, together with the standardized ratio scale described in Mueller Loose and Lockshin (2013), which allows us to identify the importance of each attribute with respect to the most important attribute.

In this study, each of the attributes employed for the BWS analysis was presented a maximum of three times with a distinct choice in the questionnaires employed. Thus, the individual BWS scores range from −3 to +3. In a similar pattern, as shown in Table 4, for the total sample, the market access attribute carries 89.2% importance relative to the economic benefits attribute; additional services and control mechanisms carry 70.7%, cooperative ideology 67.9%, the governance model 66.6%, and finally communication with 67.9%.

The analysis of variance shows that there are no significant differences across the samples' regional sub-groups, since the p-value for all seven attributes is found to be more than 0.001 (0.933). However, further ranking of the cooperative attributes across the different regions is found to be unique for each region separately. Economic benefits are found to be ranked 1st for both the Zagora and Pouri segments while for Makryrachi it was ranked 3rd and less important than the cooperative ideology and market access attributes. Furthermore, cooperative ideology which ranked as the most important (1st) for the Makryrachi segment is found to be the least important (7th) for the two other regional sub-groups of the sample. The high importance attached to the cooperative ideology and market access attributes by members established in Makryrachi could be related to the fact that the Makryrachi farmers recently joined the ZAGORIN cooperative and have a need for secured markets and solidarity in their economic activity. The market access attribute is found to be equally important for all three regional sub-groups as it was ranked 2nd for Zagora and Makryrachi and 3rd for Pouri, while the attribute related to the quality of communication in the cooperative was ranked 5th, 4th, and 6th between the seven attributes for Zagora, Makryrachi, and Pouri, respectively.

The identified high importance that cooperative members attach to the provision of economic benefits (0.62) and market access (0.32) are presented in Table 4. The importance that members attach to economic benefits and market access comes in accordance with the results of the relative cooperative literature (Fulton, 1999; Van

Table 4 Average individual-level BWS scores of cooperative attributes across different sub-groups of the sample and analysis of variance

	Total		Zagora			Makryrachi			Pouri			F/ Chi-square	p-value
Cooperative attribute	Score	Ratio[a]	Score	Ratio	Rank	Score	Ratio	Rank	Score	Ratio	Rank		
Economic benefits	0.62	100.0	0.51	100.0	1	0.25	62.5	3	1.53	100.0	1	343	0.933
Additional services	−0.12	70.7	−0.08	76.2	3	−0.42	45.6	6	−0.13	45.0	4	4.33	0.933
Market access	0.32	89.2	0.35	95.6	2	0.33	64.5	2	0.13	52.5	3	209.33	0.933
communication	−0.24	66.2	−0.18	72.0	5	−0.17	51.0	4	−0.60	34.2	6	36.33	0.933
Cooperative ideology	−0.22	67.9	−0.25	70.3	7	0.92	100.0	1	−0.93	31.2	7	270.33	0.933
Cooperative governance model	−0.24	66.6	−0.14	73.8	4	−0.58	38.0	7	−0.47	38.4	5	5.33	0.933
Control mechanisms	−0.13	70.7	−0.22	71.5	6	−0.33	47.9	5	0.47	61.8	2	144.33	0.933

[a] "Ratio" refers to the standardized ratio scale as described in Mueller et al. (2010)

Dijk, 1999; Cook & Chaddad, 2004; Kalogeras et al., 2007; Nishi & Kumar, 2011; Liebrand & Ling, 2014). The results of the BWS analysis suggest that the cooperative members of ZAGORIN attach lower importance to communication (−0.24), the cooperative's governance model (−0.24), and cooperative ideology (−0.22). These results are not in line with several studies in the literature about cooperatives (Trechter et al., 2002; Bijman & Verhees, 2011; Flecha & Ngai, 2014; Iliopoulos & Theodorakopoulou, 2014; Morfi et al., 2015; Verhees et al., 2015; Van Dijk et al., 2019), which suggest that those attributes are crucial for motivating members to remain committed to their cooperative. It may be that context specificity (e.g., type and size of cooperative, cross-cultural differences, structure of the product/service market) should be accounted for when eliciting member preferences for cooperative attributes.

4.2 LCA Results

Furthermore, the classification of the sample was carried out by employing a latent class cluster analysis (LCA) using the STATA MP/16 software. To fulfill the LCA, we utilized the individual BWS scores of the participants. According to Weller et al. (2020) and Chrysochou et al. (2022), several statistical criteria have been offered to evaluate which class model is the best fit according to different datasets. However, the choice of the resulting best fit class models should always be accounted for with respect to interpretability.

Table 5 reports two criteria that were taken into consideration for the assessment and final choice of the best fitting model: the Akaike information criterion (AIC) and, more importantly, the Bayesian information criterion based on the log-likelihood (BIC_{LL}). The decision to choose the most accurate solution model was based on the value of the BIC_{LL} being small compared to corresponding values for the other solution models (Weller et al., 2020). The latent two-class model satisfied the above-mentioned criteria. Furthermore, the two-class solution model was considered the most interpretable regarding the evidence it provided, and was thus retained as the most appropriate among the five solution models.

Table 5 Latent class cluster models based on best-worst importance scores

Model	AIC	BIC_{LL}
One-cluster model (independence)	2870.027	2907.315
Two-cluster model	2829.917	2888.513
Three-cluster model	2811.406	2891.309
Four-cluster model	2803.461	2904.672
Five-cluster model	2804.746	2927.264

4.2.1 LCA: BWS Individual-Level Scores

The two segments were labeled and interpreted based on the dominant scores within each segment, which are presented in Table 6. The first segment was labeled as "business-oriented" and is the largest among the two; it contains 69.9% of the total sample of respondents. This particular segment rates the economic benefits of a cooperative's operations as the most important cooperative attribute regarding their motivation to remain committed (1.73). The other cooperative attribute that is found to be important for business-oriented participants is the market access cooperative attribute (0.41). On the opposite side of the spectrum, the remaining cooperative attributes presented in this study were evaluated as less important for the business-oriented class. The least important was found to be cooperative ideology (−0.77), followed by communication and the cooperative's governance model (−0.53), additional services (−0.23), and control mechanisms (−0.1).

The second segment, labeled as "ideology seekers," contains 30.1% of the participants. Compared to the other segment, this segment scores highest on cooperative ideology (0.94), followed by the cooperative's governance model (0.39) and communication attributes (0.38), market access (0.12), and additional services (0.1). The least important attribute for this segment was found to be the economic benefits (−1.73) and the control mechanisms (−0.21).

The results of the latent class analysis revealed two segments relative to latent variables. The results presented in Table 6 indicate that the sample is clearly splintered into two different sub-groups, and the main factor for defining the difference between them is found to be in the importance that a member attaches to the provision of economic benefits. Evidently, the aggregate individual BWS score for the

Table 6 Mean scores, differences, and segment size for the two-cluster model across location regions

Cooperative attribute	Class 1 "Business-oriented"	Class 2 "Ideology seekers"	F/ Chi-square	*p*-value
Economic benefits	1.73	−1.73	98.23	<0.001
Additional services	−0.23	0.1	5	0.543
Market access	0.41	0.12	6.81	0.339
Communication	−0.53	0.38	13.18	0.036
Cooperative ideology	−0.77	0.94	291.83	<0.001
Cooperative governance model	−0.53	0.39	13.3	0.021
Control mechanisms	−0.1	−0.21	7.81	0.252
		Segment size		
N(=106)	73	33		
Total (%)	68.9	31.1		
Zagora (%)	65.8	34.2		
Makryrachi (%)	58.3	41.7		
Pouri (%)	93.3	6.7		

business-oriented segment is found to be the highest among the other attributes (1.73), while the corresponding value for the ideology seekers segment is found to be the lowest among the others (−1.73); this could be related to the differentiation in the cooperative's business orientation between the two segments. On the one hand, we have the members that prioritize the economic compensation when doing business with the cooperative, while on the other hand, we have the members that prioritize the bonds of solidarity and social promotion in the cooperative business. Consequently, other than the economic benefits, the results indicate a difference in the perceived importance of cooperative ideology between the two distinct subgroups. Other than the cooperative ideology, the ideology seekers are found to highly evaluate the quality of communication and effectiveness of the cooperative's governance-related attributes.

4.2.2 LCA: Socio-demographic Differences

Table 7 presents results regarding the socio-demographic differences of the participants relative to the different classes identified through LCA. Regarding gender, for both segments: "business-oriented" and "ideology seekers," the male population is present at a higher rate in an equally distributed manner for business-oriented (69.9%) and ideology seekers (69.7%), respectively, while the corresponding numbers for women are 30.1% and 30.3%. Regarding age, the business-oriented participants appear to be of a younger age since the percentage of members less than 29 years old is found to be 20.55%, while the corresponding percentage for ideology seekers is 9.09% of the total segment population. In the same manner, the percentages of members being more than 65 years of age are found to be 12.12%, while there were no members older than 65 years of age in the business-oriented segment. Regarding education, the business-oriented segment appears to have a slightly higher proportion of members with a lower education level compared to the ideology seekers; members with a level of education above high school in the ideology seekers segment was found to be 39.39% of the total population, while the corresponding percentage for the business-oriented segment was found to be 30.1%.

Regarding the monthly family income, the ideology seekers segment appears to have a slightly higher level of monthly earned income, since 57.6% of respondents answered that their family earns more than 1000€ per month, while on the other hand, 58.9% of the business-oriented segment reported that they earn less than 1000€ monthly. Regarding the members that constitute the household of each respondent, 51.5% of the ideology seekers reported living in a household consisting of two or less members, while 70.8% of members belonging to the business-oriented segment reported that their household comprises at least three members. Finally, regarding the area of the land bearing fruits, answers reveal that a large portion of members owns 30,000 m^2 or less of land bearing fruits, specifically 79.5% for the business-oriented members and 75.7% of the ideology seekers, respectively.

However, the p-values provided by the chi-square test regarding the association of the socio-demographic characteristics of the latent segments for the differences

Table 7 Socio-demographic differences across latent class segments

	Class 1 "Business-oriented"	Class 2 "Ideology seekers"	F/ Chi-square	*p*-value
Gender (%)			0	0.986
Male	69.86%	69.70%		
Female	30.14%	30.30%		
Age groups (%)			2.04	0.052
Less than 20	1.37%	0.00%		
20–29	19.18%	9.09%		
30–39	28.77%	24.24%		
40–49	16.44%	18.18%		
50–65	34.25%	36.36%		
More than 65	0.00%	12.12%		
Education (%)			2.69	0.843
High school or below	69.86	57.58		
Bachelor's degree	26.03	33.33		
Master's degree or above	4.11	6.06		
Family income (%)			2.69	0.61
Less than 500 €	6.85%	6.06%		
500–1000 €	52.05%	36.36%		
1000–2000 €	24.66%	36.36%		
2000–3000 €	8.22%	9.09%		
More than 3000 €	8.22%	12.12%		
Household members (%)			9.67	0.085
1 member	1.37%	9.09%		
2 members	17.81%	42.42%		
3 members	32.88%	27.27%		
4 members	31.51%	21.21%		
5 members	6.85%	0.00%		
More than 5 members	2.74%	0.00%		
Area of land bearing fruits (in 1000 m^2) (%)			0.86	0.930
0–15	31.51%	33.33%		
16–30	47.95%	42.42%		
31–45	8.22%	6.06%		
46–60	8.22%	12.12%		
More than 60	4.11%	6.06%		

Note: The percentages that appear below each segment are relative to the size of the total segments' population and not the total sample of this study

between the two classes were found to be higher than 0.001 in their totality. Thus, the very large *p*-values suggest that there is no association among the socio-demographic differences between the two classes, and the identified results are most probably attached randomly.

5 Conclusions

Overall, our analysis, which is based on a member-segmentation solution, identifies two segments in our sample: the business-oriented and ideology seekers sub-groups. These member segments differ regarding the way their members prioritize their preferences for seven cooperative attributes. This study fills the gap in the relevant literature, which has been regarded not as measuring the commitment of members in a cooperative, but on evaluating members' preferences for the attributes of a sustainable multi-purpose cooperative established and operating for more than 100 years, the ZAGORIN cooperative, and identifying the diversity in their preference structure. The utilization of the BWS and LCA analyses could answer this research objective, as they approach the members' preferences for core cooperative attributes in a holistic manner.

Some useful managerial implications can be derived from this study. Firstly, the findings highlight the emergence of two segments of cooperative members regarding their motives for commitment; thus, an opportunity for forming a framework for cooperative strategies aiming at enhancing commitment is provided, subject to the members' identified preferences of attributes affecting their sense of commitment. Secondly, the fact that the largest segment of members (68.9%) prioritizes economic benefits for their commitment to the collective enterprise highlights the significance of the efforts made toward the stabilization of the annual financial returns that the ZAGORIN cooperative has managed to accomplish for their members. Finally, the existence of a small segment that prioritizes cooperative ideology, communication, and the cooperative governance model suggests that there are a number of cooperative members that are fully aware of the cooperative's "intangible" value. Thus, there is the potential for acting toward the enhancement of cooperative ideology and strengthening of communication with non-aware members in order to establish a uniform member base that fully understands the cooperative as his/her own business and not just a marketing channel through which he/she markets their produce.

Regarding the enhancement of the level of commitment of member farmers, we can suggest that the cooperatives in Greece may implement strategies toward two key directions. Firstly, as highlighted by the revealed member preferences of the niche sample segment of "ideology seekers," the cooperatives should lay emphasis on increasing the quality of communication mechanisms with their members in order to build a stable environment characterized by strong trust bonds between all agents of the cooperative enterprise. Good quality communication between the cooperative management and the members will inspire the latter to show more understanding in times of a cooperative's financial depression, and the former to fully understand the needs of their members. The implementation of educational and training projects for informing member and non-member farmers regarding the unification of small-scale farmers under the values of cooperation can be used as a tool for addressing the increased competition of the globalized food markets.

Secondly, regarding the coop management, by monitoring the high levels of commitment resulting from the very well-established and successful marketing operations of the ZAGORIN cooperative, we draw the following suggestion regarding the enhancement of commitment in Greek agricultural cooperative institutions. Cooperatives should invest in hiring specialized employees that will assist in organizing and operating the managerial, marketing, and financial domains of the collective enterprise. As supported by the BWS data regarding the relative importance of market access as a cooperative attribute, cooperatives should enhance their business operations and techniques of communication, by increasing the value of the cooperative as a business whose member-investors will perceive their membership as a high importance asset. In this sense, their commitment will be strengthened and they will be more willing to further invest more in the collective enterprise. In addition, a well-organized and proper functioning endowment will certainly motivate the members to participate more actively in the collective affairs as well. However, the employment of external employees is going to increase the operating costs of the cooperative, but in times of increased competition, it appears to be a necessity, and this market-based asset will have a positive impact on a cooperative's outcomes in the long run.

6 Limitations and Future Research

Since our study was focused on one medium-size (≈700 members) cooperative, further empirical studies need to be employed to examine the relative importance of the identified cooperative attributes in larger populations. Similar studies may be conducted in cooperatives operating in differentiated frameworks to make a comparison between cross-cultural cooperative organization members. Context specificity may be an important factor to account for future studies. Additionally, further empirical research is required toward the more informative identification of the cooperative attributes that drive the members' utility and ultimately result in their commitment to the collective enterprise. The findings of studies similar to the current one, regarding the domestic cooperative sector, would inform practical responses to the low commitment problem of Greek cooperatives and indicate whether the tentative suggestions we have made above can be useful in differentiated cultural contexts.

Finally, the BWS seems to be an appropriate method for evaluating the preferences of a given group on several attributes. Nevertheless, there are limitations related to its methodological approach. The most important is the fact that BWS identifies relative and not absolute importance (Mueller Loose & Lockshin, 2013). In addition, we encountered limitations during the data collection procedure. Those limitations are inextricably bound with the implementation of the BWS and are related to the extent that the participants found the procedure confusing (Cohen & Orme, 2004). This might be due to the time needed to be spent on behalf of the respondent, in order to fully understand the procedure of answering the questionnaire.

References

Abrisham, A. (2011). Barriers of local participation in rural cooperatives in Fars Province, Iran. *Journal of American Science, 7*(1), 670–673.

Avgeris, A., Kontogeorgos, A., & Sergaki, P. (2018). The "Reciprocity" game: A theoretical basis for measuring reciprocity in human socio-economic interactions. *International Journal of Social Sciences, 7*(1), 13–33.

Arcas-Lario, N., Martín-Ugedo, J. F., & Mínguez-Vera, A. (2014). Farmers' satisfaction with fresh fruit and vegetable marketing Spanish cooperatives: An explanation from agency theory. *International Food and Agribusiness Management Review, 17*(1), 127–146.

Barraud-Didier, V., Henninger, M. C., & El Akremi, A. (2012). The Relationship between members' trust and participation in the governance of cooperatives: The Role of organizational commitment. *International Food and Agribusiness Management Review, 1*(15), 1–24.

Bhuyan, S. (2007). The 'people' factor in cooperatives: An analysis of members' attitudes and behaviour. *Canadian Journal of Agricultural Economics, 55*, 275–298.

Bijman, J., & Verhees, F. (2011). *Member or customer? Farmer commitment to supply cooperatives*. Paper presented in the International Conference on the Economics and Management of Network (EMNet), December 1–3, 2011, Limassol, Cyprus.

Bijman, J., & Ruben, R. (2005). *Repositioning agricultural cooperatives in the North and the South: Where do the Twain meet?* Presented in Agri-profocus (Agriprofocus Expert Meeting, Deventer.

Boned, O., & Mutuel, C. (2006). *Annual report*. Europe Rabobank, ICBA, 17.

Borgen, S. O. (2001). Identification as a trust-generating mechanism in cooperatives. *Annals of Public and Cooperative Economics, 72*(2), 209–228.

Chloupkova, J., Svendsen, G. L. H., & Svendsen, G. T. (2003). Building and destroying social capital: The case of cooperative movements in Denmark and Poland. *Agriculture and Human Values, 20*, 241–252.

Chrysochou, P., Corsi, A. M., & Krystallis, A. (2012a). What drives Greek consumer preferences for cask wine? *British Food Journal, 114*(8), 1072–1084.

Chrysochou, P., Krystallis, A., Mocanu, A., & Lewis, R. L. (2012b). Generation Y preferences for wine: an exploratory study of the US market applying the best-worst scaling. *British Food Journal, 114*(4), 516–528.

Chrysochou, P., Tiganis, A., Trabelsi Trigui, I., & Grunert, G. K. (2022). A cross-cultural study on consumer preferences for olive oil. *Food Quality and Preference, 97*, 1–8.

Cohen, S., & Orme, B. (2004). What's your preference? *Marketing Research, 16*(2), 32–37.

Cook, L. M., & Chaddad, R. F. (2004). Understanding new cooperative models: An ownership-control rights typology. *Review of Agricultural Economics, 26*(3), 348–360.

Dakurah, H. A., Goddard, E., & Osuteye, N. (2005). *Attitudes towards and satisfaction with cooperatives in Alberta: A survey analysis*. Presented in the American Agricultural Economics Association Annual Meeting, Providence, Rhode Island, 24–27.

Finn, A., & Louviere, J. J. (1992). Determining the appropriate response to evidence of public concern: The case of food safety. *Journal of Public Policy and Marketing, 11*, 12–25.

Flecha, R., & Ngai, P. (2014). The challenge for Mondragon: Searching for the cooperative values in times of internationalization. *Organization, 21*(5), 666–682.

Franke, T. M., Ho, T., & Christie, C. A. (2012). The Chi-Square test: Often used and more often misinterpreted. *American Journal of Evaluation, 33*(3), 448–458.

Fowler, J. H., & Christakis, N. A. (2010). Cooperative behavior cascades in human social networks. *PNAS, 107*, 5334–5338.

Fulton, M. E. (1999). Cooperatives and member commitment. *The Finnish Journal of Business Economics, 4*, 418–437.

Fulton, J. R., & Adamowicz, W. (1993). Factors that influence the commitment of members to their cooperative organization. *Journal of Agricultural Cooperation, 4*(8), 39–53.

Fulton, M. E., & Giannakas, K. (2001). Organizational commitment in a mixed oligopoly: Agricultural co-operatives and investor-owned firms. *American Journal of Agricultural Economics, 83*, 1258–1265.

Grashuis, J. (2018). An exploratory study of ownership and governance interrelationships in traditional and hybrid farmer cooperatives. *Managerial and Decision Economics, 39*(6), 664–673.

Grashuis, J., & Cook, L. M. (2018). A structural equation model of cooperative member satisfaction and long-term commitment. *International Food and Agribusiness Management Review, 22*(2), 247–263.

Gray, T. W., & Kraenzle, C. A. (1998). *Member participation in agricultural cooperatives: A regression and scale analysis*. United States, U. States, & Rural Business/Cooperative Service.

Grymshi, D., Crespo-Cebada, E., Elghannam, A., Mesías, F. J., & Díaz-Caro, C. (2021). Understanding consumer attitudes towards ecolabeled food products: A latent class analysis regarding their purchasing motivations. *Agribusiness, 38*(1), 93–107.

Hakelius, K. (1996). *Cooperative values: Farmers' cooperatives in the minds of farmers*. PhD dissertation, Swedish University of Agricultural Sciences.

Henry, K., & Muthén, B. (2010). Multilevel latent class analysis: An application of adolescent smoking typologies with individual and contextual predictors. *Structural Equation Modeling: A Multidisciplinary Journal, 17*, 193–215.

Hernández-Espallardo, M., Arcas-Lario, N., & Marcos-Matás, G. (2012). Farmers' satisfaction and intention to continue membership in agricultural marketing co-operatives: Neoclassical versus transaction cost considerations. *European Review of Agricultural Economics, 40*(2), 239–260.

Iliopoulos, C., & Theodorakopoulou, I. (2014). Mandatory cooperatives and the free rider problem: The case of santo wines in Santorini, Greece. *Annals of Public and Cooperative Economics, 85*(4), 663–681.

Isimwe, M. A., Mbabazize, P. M., & Shukla, J. (2016). Assessment on the influence of coffee cooperatives activities on member's welfare: A case study of Cocamu Cooperative, Rwanda. *The International Journal of Business & Management, 4*(4), 406–417.

Kalogeras, N., Pennings, J. M. E., van Dijk, G., & van der Lans, I. A. (2007). The structure of marketing cooperative: A members' perspective. In K. Karantininis & J. Nilsson (Eds.), *Vertical markets and cooperative hierarchies* (pp. 73–92). Springer Academic Publications.

Kalogeras, N., Pennings, J. M. E., Van der Lans, I. A., Garcia, P., & Van Dijk, G. (2009). Understanding heterogeneous preferences of cooperative members. *Agribusiness, 25*(1), 90–111.

Kollock, P. (2009). *The emergence of exchange structures: An experimental study of uncertainty, commitment and trust* (pp. 170–206). Organizational trust. R.M. Kramer, dir. Oxford University Press.

Kontogeorgos, A., Sergaki, P., & Chatzitheodoridis, F. (2017). An assessment of new farmers' perceptions about agricultural cooperatives. *Journal of Development Entrepreneurship, 22*(1).

Lee, J. A., Soutar, G., & Louviere, J. (2008). The best-worst scaling approach: An alternative to Schwartz's values survey. *Journal of Personality Assessment, 90*(4), 335–347.

Liang, Q., Huang, Z., Lu, H., & Wang, X. (2015). Social capital, member participation, and cooperative performance: Evidence from China's Zhejiang. *International Food and Agribusiness Management Review, 18*(1), 49–77.

Liebrand, B. C., & Ling, K. C. (2014). *Member satisfaction with their cooperatives: Insights from dairy farmers, USDA Rural Development*. Rural Business Cooperative Center, RBS Research Report 229.

Louviere, J., Flynn, T., & Marley, A. (2015). *Best-worst scaling: Theory, methods and applications. Handbook of choice modeling* (pp. 178–201). Cambridge University Press.

Manousakis, T., Sergaki, P., & Van Dijk, G. (2021). Identification of strategies for enhancing member commitment in Greek agricultural cooperatives: The opinion of cooperative experts. *Revista Nacional de Administración, 12*(2). https://doi.org/10.22458/rna.v12i2.3777

Marley, A. J., & Louviere, J. (2005). Some probabilistic models of best, worst, and best-worst choices. *Journal of Mathematical Psychology, 49*(6), 464–480.

Morfi, C., Olilla, P., Nilsson, J., Feng, L., & Karantininis, K. (2015). Motivation behind members' loyalty to agricultural cooperatives. In J. Windsperger, G. Cliquet, T. Ehrmann, & G. Hendrikse (Eds.), *Interfirm networks* (pp. 173–190). Springer.

Mueller, S., Lockshin, L., & Louviere, J. (2010). What you see may not be what you get: Asking consumers what matters may not reflect what they choose. *Marketing Letters, 21*(4), 335–350.

Mueller Loose, S., & Lockshin, L. (2013). Testing the robustness of best worst scaling for cross-national segmentation with different numbers of choice sets. *Food Quality and Preference, 27*(2), 230–242.

Nishi, A. K. S., & Kumar, R. (2011). Dairy farmers' satisfaction with dairy cooperative societies: A case study. *Indian Research Journal of Extension Education, 11*(1).

Novkovic, S. (2008). Defining the co-operative difference. *The Journal of Socio-Economics, 37*, 2168–2177.

Österberg, P., & Nilsson, J. (2009). Members' perception of their participation in the governance of cooperatives: The key to trust and commitment in agricultural cooperatives. *Agribusiness, 25*(2), 181–197.

Sergaki, P., Kontogeorgos, A., Kalogeras, N., & Van Dijk, G. (2020). Reciprocity and cooperative's performance. The example of mandatory Greek cooperatives. *AGER: Revista de Estudios sobre Despoblación y Desarrollo Rural (Journal of Depopulation and Rural Development Studies), 29*, 7–38.

Staatz, J. M. (1987). The structural characteristics of farmer cooperatives and their behavioral consequences. In J. S. Royer (Ed.), Cooperative theory: New approaches. ACS Service Rep. 18, July. US Department of Agriculture.

Trechter, D. D., King, R. P., & Walsh, L. (2002). Using Communications to influence member commitment in cooperatives. *Journal of Cooperatives, 17*, 1–19.

Valentinov, V. (2004). Toward a social capital theory of cooperative organization. *Journal of Cooperative Studies, 37*(3), 5–20.

Van Dijk, G. (1999). Evolution of business structure and entrepreneurship of cooperatives in the horti- and agribusiness. *The Finish Journal of Business Economics, 4*, 471–483.

Van Dijk, G., Sergaki, P., & Baourakis, G. (2019). *The cooperative enterprise: Practical evidence for a theory of cooperative entrepreneurship*. Springer.

Verhees, J. H. M., Sergaki, P., & Van Dijk, G. (2015). Building up active membership in cooperatives. *New Medit, 14*(1), 42–52.

Vermunt, J. K., & Magidson, J. (2002). Latent class cluster analysis. In J. A. Hagenaars & A. L. McCutcheon (Eds.), *Applied latent class analysis* (pp. 89–106). Cambridge University Press.

Weller, B. E., Bowen, N. K., & Faubert, S. J. (2020). Latent class analysis: A guide to best practice. *Journal of Black Psychology, 46*(4), 287–311.

Consumption Patterns for Healthy and Environmentally Friendly Food Choices: An Overview of Contemporary Issues

Dimitra Lazaridou, Konstadinos Mattas, Efthimia Tsakiridou, and Murat Yercan

1 Introduction

In recent years, new trends in consumer buying behavior emerge. Buyers are looking for less-processed products and become more desirous of a better quality of life. Many consumers believe in the motto "You are what you eat" and they make their food choices accordingly. It has also been ingrained in societies that food consumption has a large environmental impact. However, this impact can be reduced substantially by changing consumers' food-related behaviors. Both consumption trends for healthier food and consumers' consciousness to mitigate the environmental impact of their choices have led to the increase in demand for environmentally friendly products.

Organic food encompasses natural food items which are free from chemicals, fertilizers, pesticides, etc. On account of being produced without chemicals, organic food is considered an environmentally friendly food choice. Additionally, the health benefits were perceived as an important attribute of organic food. It is regarded as more nutritious than conventional food. At the same time, turning to

D. Lazaridou (✉)
School of Agriculture, Department of Agricultural Economics, Aristotle University of Thessaloniki, Thessaloniki, Greece

Department of Forestry and Natural Environment Management, Agricultural University of Athens, Karpenisi, Greece
e-mail: dlazaridou@aua.gr

K. Mattas · E. Tsakiridou
School of Agriculture, Department of Agricultural Economics, Aristotle University of Thessaloniki, Thessaloniki, Greece

M. Yercan
Faculty of Agriculture, Department of Agricultural Economics, Ege University, Bornova, Izmir, Turkey

K. Mattas et al. (eds.), *Value Chain Dynamics in a Biodiverse Environment*, Cooperative Management, https://doi.org/10.1007/978-3-031-49845-9_4

environmentally friendly food choices reduces the food-related carbon footprint and contributes to climate change mitigation. Organic food is gaining acceptance not only in Europe and North America, but also in developing countries such as China and India (Paul et al., 2016).

These new dietary patterns contributed to farmers turning to more efficient and environmentally friendly food production techniques. Increasing consumers' demand for environmentally friendly food augments the rate of organic farming implementation and decreases the level of farmers' risk. From the farmers' point of view, it is important to receive information on consumer demand for organics to support farming decisions. Consumers' attitude is an important predictor of their intention to buy organic food. From the review of the literature, there arises a wide breadth of papers covering subjects related to consumers' food habits and attitudes toward adopting environmentally friendly food consumption. There are also papers examining the effects of biodiversity conservation on consumer purchase decisions.

The aim of this study is to provide a critical review of the trend toward healthy and environmentally friendly food. In this framework, this chapter does the following: (1) it provides an overview of the scholarly literature on the consumption patterns shaped by buyers' attitudes toward healthy and environmentally friendly products; (2) it highlights the research findings about consumers willing to change their food choices in a more environmentally friendly direction; and (3) it demonstrates the existing policy recommendations that reinforce public awareness and encourage environmentally friendly food consumption patterns.

2 Methodology

To address the objective of this chapter, various steps were followed for selecting the related literature. A systematic literature search was used to identify relevant works on the baseline aim. A search was carried out on the Scopus bibliographic database (URL http://www.scopus.com) on June 2, 2022. The terms "Consumption patterns" **AND** "Environmentally friendly foods" **OR** "Healthy foods" were entered in the search fields "Article title, Abstracts, Keywords." All studies derived from the Scopus database (n = 138) were stored. Access was not allowed to four papers (n = 4), and therefore, they were not examined. Review papers (n = 12), an erratum publication (n = 1), and publications on languages other than English (n = 8) were also excluded. Studies were not considered relevant, as they did not comprehensively explain issues related to consumption patterns and healthy or environmentally friendly food choices were omitted (n = 72) as well. Forty-one articles (n = 41) ware read carefully and analyzed in depth for this chapter. An overview of the methodological framework is presented in Fig. 1.

We reviewed and interpreted those research articles, which contributed to our understanding about consumers' attitude and behavior toward healthy and environmentally friendly food choices. Finally, some additional research papers were

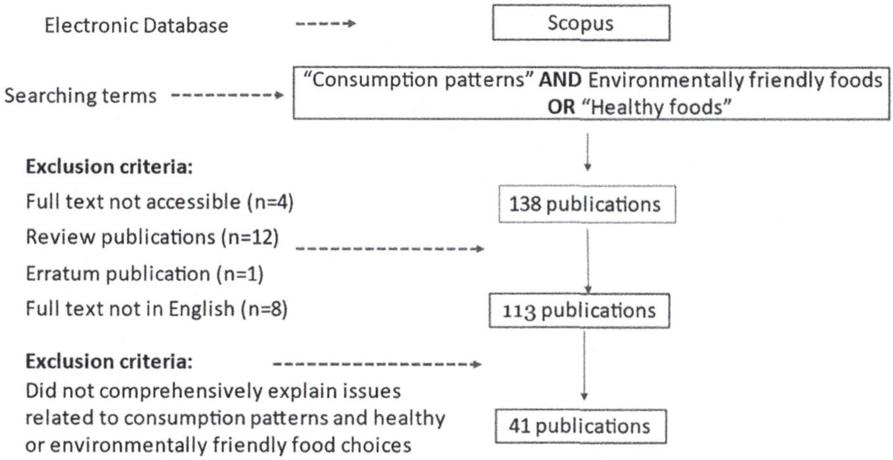

Fig. 1 Flow chart of successive stages in the literature review process. (Date of literature search: 2 June 2022)

Fig. 2 Overview of studies by year of publication (the literature review covers results until June 2, 2022)

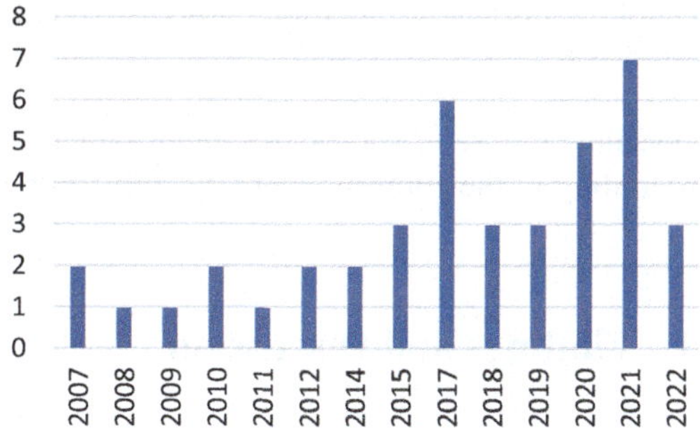

examined to better understand how consumer needs on healthy products shape the consumption patterns or cultivation trends.

3 Descriptive Analysis

The descriptive analysis reveals that the research for "Consumption patterns with an assessment of consumer need for environmentally friendly and healthy foods" is of constant interest over time. It also presents a stronger appearance of publications in the year 2021. An overview of the studies by year of publication is presented in Fig. 2.

From the geographic scope of the selected studies, the European continent has a dominant presence accounting for about 55%, followed by Asia, America, and

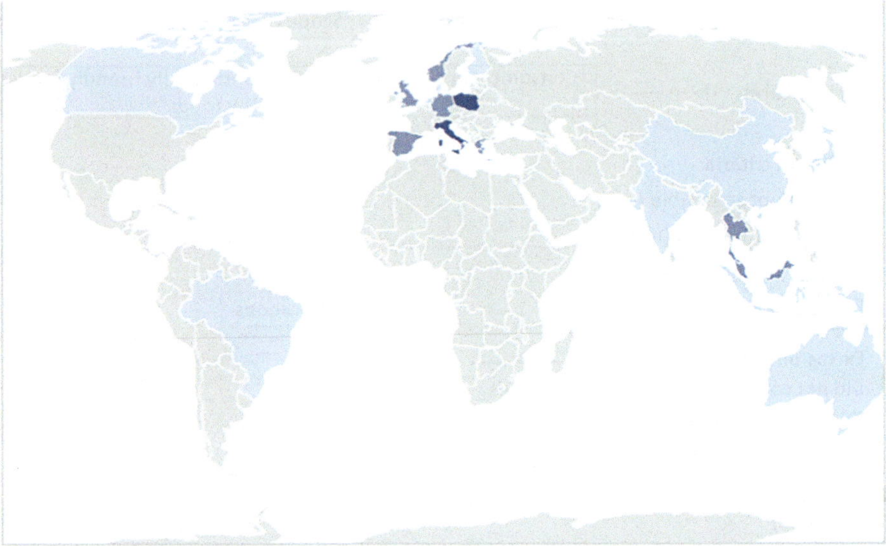

Fig. 3 Distribution of the papers analyzed according to where the countries' surveys were performed

Oceania. Figure 3 shows the geographical distribution of the studies analyzed, according to the country where the corresponding surveys were performed.

4 Findings and Discussion

Many studies in the literature have discussed several factors which affect the demand for organic foods, as they are by far the most successful environmentally or ethically "enhanced" type of product worldwide (Thøgersen et al., 2015). Some studies highlighted the health concerns and the perceived higher nutritional value of organic food as the main drivers for their demand (de Magistris and Garcia, 2008; Michaelidou & Hassan, 2008; Salleh et al., 2010; Kushwah et al., 2019). Other studies cited concerns for environment and climate change as important demand drivers (Nuttavuthisit & Thøgersen, 2017; Ghali, 2019). According to other findings, socioeconomic factors can be considered as the best predictor of consumer behavior toward environmentally friendly foods, whereas there are studies that indicate human character factors to contribute toward influencing a consumer's purchase intention for healthy and environmentally friendly food. Some studies imply that consumers are willing to pay more for risk-free food (Güney & Sangün, 2017; Mazzocchi et al., 2019; Karelakis et al., 2020), whereas others conclude that a conventional consumer is not completely prepared and willing to pay a premium price for a sustainable food product (De Daverio et al., 2021).

4.1 *Factors Influencing Consumer Attitudes and Habits for Environmentally Friendly Food*

4.1.1 Socioeconomic Status

Many researchers have examined and outlined the crucial factors that have an influence on consumers' decisions for healthy and environmentally friendly food. Regarding the food habits, they are profoundly affected by social and economic factors. In particular, trying to identify consumers' attitudes and behavior toward organics, Tsakiridou et al. (2008) carried out a questionnaire survey to analyze the demographic variables highly correlated with organics' attitudes and consumption. Higher education levels and higher income indicated a strong correlation with their consumption. In addition, environmental and health concerns are factors that affect preferences for organics' consumption. The survey was carried out by Welsch and Kühling (2009) who explored the primary factors behind the shift in the consumer attitude toward environmentally friendly food. By applying a questionnaire survey, they identified as favorable factors for purchasing organic food the pro-environmental attitudes (protective actions toward the environment), the acknowledgment that environmental problems are exaggerated, and the shift toward renewable energy sources.

Recognizing that the demand for food has been growing due to population increase, Ariani et al. (2021) carried out a survey to analyze current and eco-friendly household consumption. For the synthesis of their technical review, secondary data and information about household food waste were collected from Statistics Indonesia and Indonesian Agency for Food Security. According to their findings, driving factors of household food choices were based on social, economic, and cultural aspects rather than environmental considerations. Concerning the economic variables, results suggested that the higher the income, the better the quantity and quality of the food consumed. Regarding the food waste produced by each household, it seems that the larger proportion came from vegetables and fruits. In the same spirit, Esteve-Llorens et al. (2021a) in their survey attempted to examine the influence of social, economic, and cultural aspects on household food choices. They explore variations in food consumption patterns in terms of GHG emissions and nutritional intake adequacy for different climatic zones in Spain. Their results show that the daily food basket and eating habits, associated with different territories (climatic zones), were justified on the basis of different culinary culture and traditions, economic levels, and socio-demographic profiles. The higher carbon footprint recorded in some regions was due to higher consumption of animal-origin products. On the other hand, consumption of higher amounts of fruits, seafood, and legumes provided some regions (the northern region of Spain) with a better nutritional profile. The perception that organic foods are more traditional and safe makes consumers value them as well, according to a transnational survey in China and Brazil (Thøgersen et al., 2015).

Additionally, the social desirability of behaving in a more environmentally friendly way can strongly motivate consumers to purchase environmentally friendly foods (Kaaronen & Strelkovskii, 2020). Emotions have a crucial role in mentoring consumers toward pro-environmental food consumption as well. Onwezen (2015) examined the function of emotions in purchasing choices for pro-environmental food. The survey was performed through questionnaires that were filled out by Dutch respondents, and it was concluded that both private and collective emotions can mentor decisions toward environmentally friendly food choices.

Raptou and Manolas (2022) conducted a survey among 807 adults via a formal questionnaire in supermarkets and food stores, examining consumption patterns for organic foods as well. Results showed that consumers who acknowledged the benefits of the organic food production system had a higher likelihood to purchase organics, on a regular basis. Moreover, consumption of organic foods was positively associated with health consciousness and climate change concerns. In another survey, Esteve-Llorens et al. (2021b) examined the food consumption patterns at the household level. They attempted to identify both the impacts that foodstuffs included in the food basket cause in the environment and the socioeconomic variables that influence the consumer choice. Data for household food consumption were collected from the Spanish Ministry of Agriculture, Fishing and Food, whereas the sample examined concerned about 12,000 households, randomly selected. The results indicated a decrease of the food-related carbon footprint over the years. However, this decrease is not always synonymous with a healthier diet for the consumers. It is also interesting that their results differ from the trend for healthy and environmentally friendly food, as they observed an increase in the consumption of processed foodstuffs and ready meals, which further distances the dietary pattern from the traditional recommendations.

4.1.2 Consumers' Environmental Friendliness

Crippa et al. (2021) calculated that food production is responsible for 34% of total GHG emissions, whereas packaging, processing, retail, transportation, consumption, and waste management have a significant contribution likewise. The production of meat is pointed out as a significant source of greenhouse emissions, while transition toward plant-based and low-meat diets has been proposed as a pathway to mitigation of climate change. In this framework, Austgulen et al. (2018) investigated whether Norwegian consumers are willing and able to change their food choices, in a more climate-friendly direction, and what factors influence their perceptions for environmental measures related to food and meat consumption. The results indicate that consumers have limited knowledge about environmental impacts of meat consumption and are unaware of its negative climate impacts. It is also highlighted that most consumers are still not ready to consume food based on what is best for the climate or environment, such as eating less meat and increasing vegetable purchases. For evaluating the effect of dietary preferences and food consumption on climate change and for quantifying their cumulative environmental impact, Churak et al. (2021) analyzed specific food consumption patterns and calculated the amounts of popular food consumed in Thailand. Their results revealed that the highest accumulated GHG emissions were identified in the group of overweight participants, noting that this group related to more animal-based food consumption. Avetisyan et al. (2014) tried to assess GHG emissions associated with consumption of domestic and imported food products, and the trade-off between production and transport emissions. Based on secondary data, they conclude that encouraging consumption to local food products reduces global GHG emissions, but only when implemented in regions with relatively low emissions intensities.

The role of London's community gardens in promoting participants' environmentally friendly food choices and habits has also been evaluated (Kim, 2017). The survey was based on (1) semi-structured interviews with community garden participants and (2) an online questionnaire survey among 48 community gardens. The findings show that the majority of gardeners surveyed tend to have lower footprints than the general population. However, they still have some carbon-intensive food consumption habits, such as consuming meat regularly and shopping in supermarkets.

4.1.3 Consumers' Health Consciousness

Consumers' healthy lifestyle is a significant predictor of consumption patterns (Djermani et al., 2021), while health consciousness has been considered as the best predictor of consumer behavior toward organic food (Michaelidou & Hassan, 2008). In particular, the health factors create a positive attitude about environmentally friendly food consumption, as a way to prevent incidence of lifestyle diseases, such as heart disorders (Shaharudin et al., 2010; Paul & Rana, 2012). Indeed, the trend for a healthier diet has generated an increasing competition on healthy food products within the food market industries. As supported, without access to healthy foods, a nutritious diet is out of reach. For this reason, numerous surveys have been carried out to investigate personal value variables and public attitudes toward healthy food consumption, in order to gain a better understanding of the healthy lifestyle consumer.

Salleh et al. (2010), through a survey conducted among 136 University lecturers in Malaysia, found that health consciousness strongly motivates consumers to purchase environmentally friendly food, while quality and taste are also significant considerations. Dean et al. (2007) utilized results from 2094 questionnaires from four countries (UK, Italy, Finland, Germany) to investigate public perceptions related to different healthy grain foods (bread, pasta, and biscuits) and examined how these perceptions are influenced by gender, nationality, and type of health claim (general vs. specific). Results confirmed that women perceived more benefit in products with general health claims (e.g., omega-3 and brain development) and men in products with specific health claims (e.g., cholesterol). Similarly, according to Rathi et al. (2017), females (girls) had more nutritious dietary intakes than boys (in India), whereas the urban adolescents reported poor dietary intakes. In this framework, Divine and Lepisto (2005) claim that people who maintain a healthy lifestyle tend to be older, female, and more educated and place less importance on the value of "excitement." Exploring the human character factors that influence individuals' choices when they make decisions on what types of food to consume, Wang and Zhang (2016) inferred that high power individuals are more likely to consume healthy food than indulgent food, while low power individuals are more likely to consume indulgent food than healthy food.

The effect of the income factor on consumer behavior is demonstrated by several papers. Healthier foods tended to be more expensive than their less healthy equivalent, in many countries. It explains why low-income households turn to less healthy food-purchasing behavior (Muzigaba & Puoane, 2011). Casini et al. (2015) investigated the evolution of the food patterns in the past decade, in order to interpret them in light of demographic characteristics and sociocultural changes. By applying latent class clustering analysis to the food spending of a sample of consumers in Italy, they tried to identify the principal food patterns. In particular, they examined the variables that were associated with the pattern of the so-called healthier consumers, who are consumers with a dietary mix in line with the recommendations of health authorities (i.e., consumers that buy fruits, vegetables, and fish). This choice gains ground among consumers with a higher level of education and among couples, but also among families with children.

4.1.4 Food Marketing and Pricing Policy

Other surveys examined the impact of food marketing and pricing policy in consumer choices for environmentally friendly food. Taghikhah et al. (2020) attempted to assess the effectiveness of different policies and informational-education campaigns to influence consumer choices, focusing their research on wine. According to their results, raising consumer awareness and increasing tax on less environmentally friendly wines turned out to be more successful in promoting organic wine. Additional findings showed that consumers greatly value the contribution of organic foods to support and strengthen the local economy and community. Therefore, they tend toward organic food in order to modify their consumption patterns for societal benefit (Thomas & Gunden, 2012). On the other hand, Sulaiman et al. (2017) examined the relationship between the consumer preferences toward healthy food, with the influence exerting the marketing mix (4P—product, place, price, and promotion). Their survey was conducted among 400 respondents, consisting of undergraduate students from the University of Utara (Malaysia). The results showed a significant and positive relationship between marketing mix and consumer preferences toward healthy food.

4.1.5 The Contribution of Eco-labeling and Advertising

The purpose of eco-labels is in redirecting consumption to more environmentally friendly food choices. They have been recognized as a means of promoting products with lower environmental impact. A considerable body of literature has focused on the effect of labels on the consumer's choice to purchase environmentally friendly foods. Some of the selected surveys are listed below.

Slapø and Karevold (2019) test different traffic-light labels to investigate their influence on consumer selection for environmentally friendly dishes. In their experimental study apart from the labels, they placed auxiliary posters to explain the labeling systems and inform consumers about the climate impact of various food categories. The results support a significant reduction on sales of meat dishes, due to the information provided. However, the sales share of vegetarian dishes and fish was not influenced. The presence of packaged food labels which carry nutritional information attracted research interest, too. According to Kozup et al. (2003), consumers have a more positive attitude toward products with detailed label descriptions and health claims. It was also demonstrated that comprehensive sustainability labels could considerably influence consumption patterns for healthy food products (Engels et al., 2010). Kraus (2015) performed a questionnaire survey among 200 respondents in Poland to determine the most important traits of functional food products attributed by consumers. Among the quality attributes highly evaluated by consumers are safe food, natural products, and healthy products. Consumers also considered significant the attributes of packaging and labeling to include information on the healthful properties of each product. However, it seems that eco-labels in foods are beneficiary not only for the environment but also for the producers' income. As noted from the survey of Chang (2012), which explored the effect of eco-labels on income distribution and income inequality, the use of eco-labels in food products increases producers' income.

In terms of the contribution of media and advertising to promote food consumption and influence consumers' perceptions or behaviors regarding healthier and more environmentally friendly food, research shows that individuals gain much of their health information from the media (Kean et al., 2012). In addition, healthy eating habits were positively associated with watching television news and having higher levels of media literacy (Kean et al., 2012).

4.1.6 Parental Contribution

Acknowledging the role of parental contribution to adopt environmentally friendly food choices by children, Halicka et al. (2021) attempted to assess the impact of sustainability issues on the behaviors of parents living with young school-aged children (in Poland). They performed a questionnaire survey among 1035 adults, revealing that parents appeared well intentioned in their motives for selecting food for their children, and they were engaged in raising their children's awareness on healthier and more environmentally friendly food consumption. It also emerged that it is commonly believed that family members should have a responsibility in teaching children for the links among food, health, and environment.

4.2 Consumer Preferences for Agricultural Products Considering the Value of Biodiversity

The protection of biodiversity has gained popularity both in consumers' opinion and in scientific debate, whereas credence attributes seem to play an important role in consumer preference formation, especially for agricultural products. There have been studies in the literature that have tried to analyze the effects of nature and biodiversity conservation on consumer purchase decisions, but they are few. Most of them suggest that consumers are sensitive toward the maintenance of the balance between biodiversity conservation and agricultural production. In particular, from the review of the literature, it can be inferred that the stated preference methods (i.e., choice experiment, contingent valuation method) are used mainly to assess consumers' preferences for buying foods produced with techniques consistent with environmental stewardship.

Indeed, most published research findings claim a significant increase in consumers' concerns regarding the impact of food production on biodiversity. Moon et al. (2002) carried out a contingent valuation approach to measure behavioral intention toward purchasing agricultural commodities produced by environmentally sound practices and to protect wilderness. The study showed that the majority of participants were willing to pay a higher premium for products cultivated respecting biodiversity conservation practices. Similarly, Khai and Yabe (2015) tried to investigate the effects of biodiversity awareness on consumers' preferences for environmentally certified rice. By applying the choice experiment method, they found that consumers were willing to pay more for 1 kg of environmentally certified rice to increase crane numbers and the biodiversity level in their area. Along the same line, Mazzocchi et al. (2019) applied the choice experiment to estimate wine consumers' willingness to pay for biodiversity conservation practices in vineyards. Their results revealed that consumers were willing to pay a premium price for wine certification that takes into account biodiversity (Mazzocchi et al., 2019). Finally, Yabe et al. (2013) applied the same technique to analyze consumer preferences related to "life brand" products that improve biodiversity. They revealed that consumers' willingness to pay for these products increased as their awareness of biodiversity conservation increased. However, the same paper revealed that they placed greater importance on their health, than on environmental conservation. These findings are consistent with a finding from Salleh et al. (2010), which shows that health consciousness factors have more impact on customer purchase intention, rather than the environmental concern. Similarly, various consumer groups have been examined by Güney and Sangün (2017) about their willingness to adopt environmentally friendly food consumption patterns. Findings showed that young consumers were willing to pay more for organic and traditionally processed olive oils.

4.3 Recommendations to Promote Healthy and Environmentally Friendly Foods

Noting that social learning about the environmental impact of specific food habits can play a key role in changing conventional consumer behaviors, several academic studies provide policy and practical recommendations to promote healthier eating habits and more environmentally friendly eating. Presenting findings from different countries, Austgulen et al. (2018) highlighted that most consumers are still not ready to make food choices based on what is best for the environment. Therefore, they suggest efficient information to be provided to consumers, as part of an environmental policy design. On the same basis, some studies imply that policy makers, together with nutritionists and agronomists, should develop a food system which balances productivity, sustainability, and community nutrition fulfillment to reinforce environmentally friendly food consumption behavior (Rathi et al., 2017; Ariani et al., 2021; Mattas et al., 2022). The role of public health programs to raise children's awareness toward healthier and more environmentally friendly food consumption practices was noted by Halicka et al. (2021). They suggest the inclusion of the principles of healthy diets and sustainable food consumption into it. According to the same authors, parental responsibility is a crucial issue too and parents should be encouraged to reinforce the awareness of children to food, health, and environment (Halicka et al., 2021).

Reviewed previous studies propose additional interventions that could increase the consumption of environmentally friendly foods. Fiscal measures, such as taxes or subsidies, can be designed to change the relative prices of healthy and unhealthy foods or nutrients. So, increasing tax on less environmentally friendly food products could be a way to promote organic products (Taghikhah et al., 2020). Similarly, eco-labeling is viewed by scientists as a way of ensuring that consumers may partially improve the eco-friendliness of food consumption. Thus, some additional actions and efforts should be designed in the direction of their greater utilization (Slapø & Karevold, 2019).

Likewise, there are many challenges to the information approach. Agencies should strive and work toward raising awareness about the benefits of sustainable food choices, through advertisements on social media, exhibitions that can share related information, etc. (Sulaiman et al., 2017). Moreover, there are findings which imply that emotions are most effective in guiding intentions toward buying organic food (Onwezen, 2015). So, emotions can be used in campaigns that focus on buying environmentally friendly foods.

5 Conclusions

Based on an extensive review of the literature, our study indicates that a wide breadth of papers covers subjects related to consumers' food habits and attitudes toward the adoption of healthy and environmentally friendly food consumption. We extract findings from various studies conducted in different countries and conclude that consumers are keen to purchase environmentally friendly foods which are quite popular not only in Europe and North America but also in developing countries. Analyzing the factors shaping consumers' preferences for sustainable foods, it is highlighted that socioeconomic status, demographic factors, culinary culture, and appreciation for traditional farming processes strongly motivate consumers to turn to more sustainable food choices.

Additional findings suggest that purchasers' concern about the quality of their life can be considered as the best predictor of their behavior toward health foods. Furthermore, the adoption of healthier eating habits can contribute to address obesity by reducing the calories in products sold at stores, to increase consumption of plant-based products, to limit the consumption of animal-origin foodstuffs, processed food, and added sugars, to reduce the food-related carbon footprint, and to mitigate climate change.

Although there are not many studies in the literature related to food consumption patterns and biodiversity conservation, most of them claim that consumers are sensitive toward the maintenance of the balance between biodiversity conservation and food production. The protection of biodiversity has gained popularity among consumers, a fact that can be exploited further, leading to more sustainable purchase decisions.

Previous studies also revealed that there are policy and practical recommendations that contribute to changing conventional consumer behaviors and promoting sustainable eating habits. So, packaging, eco-labeling, and adverting should be redesigned in order to encourage the consumption of environmentally friendly products.

Acknowledgments This research was funded by the EU H2020 BIOVALUE project [Grant Agreement No 101000499] "Fork-to-farm agent-based simulation tool augmenting BIOdiversity in the agri-food VALUE chain."

References

Ariani, M., Gantina, A., Mauludyani, A. V. R., & Suryana, A. (2021). Environmentally friendly household food consumption behavior. *IOP Conference Series: Earth and Environmental Science, 892*(1). https://doi.org/10.1088/1755-1315/892/1/012023

Austgulen, M. H., Skuland, S. E., Schjøll, A., & Alfnes, F. (2018). Consumer readiness to reduce meat consumption for the purpose of environmental sustainability: Insights from Norway. *Sustainability (Switzerland), 10*(9). https://doi.org/10.3390/su10093058

Avetisyan, M., Hertel, T., & Sampson, G. (2014). Is local food more environmentally friendly? The GHG emissions impacts of consuming imported versus domestically produced food. *Environmental and Resource Economics*. https://doi.org/10.1007/s10640-013-9706-3

Casini, L., Contini, C., Romano, C., & Scozzafava, G. (2015). Trends in food consumptions: What is happening to generation X? *British Food Journal, 117*(2), 705–718. https://doi.org/10.1108/BFJ-10-2013-0283

Chang, H. H. (2012). Does the use of eco-labels affect income distribution and income inequality of aquaculture producers in Taiwan? *Ecological Economics, 80*, 101–108. https://doi.org/10.1016/j.ecolecon.2012.05.011

Churak, P., Sranacharoenpong, K., & Mungcharoen, T. (2021). Environmental consequences related to nutritional status of Thai populations. *Journal of Public Health (Germany), 29*(4), 879–884. https://doi.org/10.1007/s10389-019-01189-8

Crippa, M., Solazzo, E., Guizzardi, D., Monforti-Ferrario, F., Tubiello, F. N., & Leip, A. (2021). Food systems are responsible for a third of global anthropogenic GHG emissions. *Nature Food, 2*, 198–209.

De Daverio, M. T. T., Mancuso, T., Peri, M., & Baldi, L. (2021). How does consumers' care for origin shape their behavioural gap for environmentally friendly products? *Sustainability (Switzerland), 13*(1), 1–19. https://doi.org/10.3390/su13010190

De Magistris, T., & Gracia, A. (2008). The decision to buy organic food products in Southern Italy. *British Food Journal, 110*(9), 929–947.

Dean, M., Shepherd, R., Arvola, A., Vassallo, M., Winkelmann, M., Claupein, E., Lähteenmäki, L., Raatsa, M. M., & Saba, A. (2007). Consumer perceptions of healthy cereal products and production methods. *Journal of Cereal Science, 46*(3), 188–196. https://doi.org/10.1016/j.jcs.2007.06.007

Djermani, F., Sulaiman, Y., & Ismail, M. Y. S. (2021). The effect of product and promotion on consumption patterns with mediation effect of healthy lifestyle. *WSEAS Transactions on Environment and Development, 17*, 520–532.

Divine, R. L., & Lepisto, L. (2005). Analysis of the healthy lifestyle consumer. *Journal of Consumer Marketing, 22*(5), 275–283.

Engels, S. V., Hansmann, R., & Scholz, R. W. (2010). Toward a sustainability label for food products: An analysis of experts' and consumers' acceptance. *Ecological Food Nutrition, 49*(1), 30–60.

Esteve-Llorens, X., Moreira, M. T., Feijoo, G., & González-García, S. (2021a). Could the economic crisis explain the reduction in the carbon footprint of food? Evidence from Spain in the last decade. *Science of the Total Environment, 755*, 142680. https://doi.org/10.1016/j.scitotenv.2020.142680

Esteve-Llorens, X., Van Dooren, C., Alvarez, M., Moreira, M. Y., Feijoo, G., & Gonzalez-García, S. (2021b). Environmental and nutritional profile of food consumption patterns in the different climatic zones of Spain. *Journal of Cleaner Production, 279*. https://doi.org/10.1016/j.jclepro.2020.123580

Ghali, Z. (2019). Motives of willingness to buy organic food under the moderating role of consumer awareness. *Journal of Scientific Research & Reports, 25*(6), 1–11.

Güney, O., & Sangün, L. (2017). Olive oil consumption attitudes as a healthy food: A survey study on young consumers. *Agro Food Industry Hi-Tech, 28*(4), 56–59.

Halicka, E., Kaczorowska, J., Rejman, K., & Szczebyło, A. (2021). Parental food choices and engagement in raising children's awareness of sustainable behaviors in urban Poland. *International Journal of Environmental Research and Public Health, 18*(6), 1–19. https://doi.org/10.3390/ijerph18063225

Kaaronen, R. O., & Strelkovskii, N. (2020). Cultural evolution of sustainable behaviors: Pro-environmental tipping points in an agent-based model. *One Earth, 2*, 85–97.

Karelakis, C., Zevgitis, P., Galanopoulos, K., & Mattas, K. (2020). Consumer trends and attitudes to functional foods. *Journal of International Food & Agribusiness Marketing, 32*(3), 266–294. https://doi.org/10.1080/08974438.2019.1599760

Kean, L. G., Prividera, L. C., Boyce, A., & Curry, T. (2012). Media use, media literacy, and African American females' food consumption patterns. *Howard Journal of Communications, 23*, 197–214.

Khai, H. V., & Yabe, M. (2015). Consumer preferences for agricultural products considering the value of biodiversity conservation in the Mekong Delta, Vietnam. *Journal for Nature Conservation, 25*, 62–71.

Kim, J. E. (2017). Fostering behaviour change to encourage low-carbon food consumption through community gardens. *International Journal of Urban Sciences, 21*(3), 364–384. https://doi.org/10.1080/12265934.2017.1314191

Kozup, J. C., Creyer, E. H., & Burton, S. (2003). Making healthful food choices: The influence of health claims and nutrition information on consumers' evaluations of packaged food products and restaurant menu items. *Journal of Marketing, 67*(2), 19–34. https://doi.org/10.1509/jmkg.67.2.19.18608

Kraus, A. (2015). Development of functional food with the participation of the consumer. Motivators for consumption of functional products. *International Journal of Consumer Studies, 39*(1), 2–11. https://doi.org/10.1111/ijcs.12144

Kushwah, S., Dhir, A., & Sagar, M. (2019). Understanding consumer resistance to the consumption of organic food. A study of ethical consumption, purchasing, and choice behaviour. *Food Quality and Preference, 77*, 1–14.

Mattas, K., Tsakiridou, E., Karelakis, C., Lazaridou, D., Gorton, M., Filipovic, M., Hubbard, C., Saidi, M., Stojkovic, D., Tocco, B., Tregear, A., & Veneziani, M. (2022). Strengthening the sustainability of European food chains through quality and procurement policies. *Trends in Food Science & Technology, 120*, 248–253. https://doi.org/10.1016/j.tifs.2021.11.021

Mazzocchi, C., Ruggeri, G., & Corsi, S. (2019). Consumers preferences for biodiversity in vineyards: A choice experiment on wine. *Wine Economic, 8*, 155–164. https://doi.org/10.1016/j.wep.2019.09.002

Michaelidou, N., & Hassan, L. M. (2008). The role of health consciousness, food safety concern and ethical identity on attitudes and intentions towards organic food. *International Journal of Consumer Studies, 32*(1), 163–170.

Moon, W., Florkowski, W. J., Bruckner, B., & Schonhof, I. (2002). Willingness to pay for environmental practices: Implications for eco-labeling. *Land Economics, 78*(1), 88–102.

Muzigaba, M., & Puoane, T. (2011). Perceived and actual cost of healthier foods versus their less healthy alternatives: A case study in a predominantly black urban township in South Africa. *East African Journal of Public Health, 8*, 278–285.

Nuttavuthisit, K., & Thøgersen, J. (2017). The importance of consumer trust for the emergence of a market for green products: The case of organic food. *Journal of Business Ethics, 140*(2), 323–337.

Onwezen, M. C. (2015). I did good, and we did bad: The impact of collective versus private emotions on pro-environmental food consumption. *Food Research International, 76*(2), 261–268. https://doi.org/10.1016/j.foodres.2015.03.032

Paul, J., & Rana, J. (2012). Consumer behavior and purchase intention for organic food. *Journal of Consumer Marketing, 29*(6), 412–422.

Paul, J., Modi, A., & Patel, J. (2016). Predicting green product consumption using theory of planned behavior and reasoned action. *Journal of Retailing and Consumer Services, 29*, 123–134.

Raptou, E., & Manolas, E. (2022). Consumption patterns and public attitudes toward organic foods: The role of climate change awareness. In W. Leal Filho & E. Manolas (Eds.), *Climate change in the Mediterranean and middle eastern region* (pp. 365–393). Springer International Publishing. https://doi.org/10.1007/978-3-030-78566-6_18

Rathi, N., Riddell, L., & Worsley, A. (2017). Food and nutrition education in private Indian secondary schools. *Health Education, 117*, 193–206.

Salleh, M., Ali, S. M., Harun, E. H., Jalil, M. A., & Shaharudin, M. R. (2010). Consumer's perception and purchase intentions towards organic food products: Exploring attitude among academician. *Canadian Social Science, 6*(6), 119–129.

Shaharudin, M. R., Pani, J. J., Mansor, W. S., & Elias, S. J. (2010). Purchase intention of organic food; perceived value overview. *Canadian Social Science, 6*, 70–79.

Slapø, H. B., & Karevold, K. I. (2019). Simple eco-labels to nudge customers toward the most environmentally friendly warm dishes: An empirical study in a cafeteria setting. *Frontiers in Sustainable Food Systems, 3*, 1–9.

Sulaiman, Y., Bakar, N. N. A. A. B., Ismail, M. Y. S., Mat, N. K. N., & Musa, R. (2017). The function of marketing mix and consumer preferences on healthy food consumption among UUM students. *International Journal of Economic Research, 14*(19), 103–122.

Taghikhah, F., Voinova, A., Shuklaa, N., & Filatova, T. (2020). Exploring consumer behavior and policy options in organic food adoption: Insights from the Australian wine sector. *Environmental Science and Policy, 109*(November 2019), 116–124. https://doi.org/10.1016/j.envsci.2020.04.001

Thøgersen, J., de Barcellos, M. D., Perin, M. G., & Zhou, Y. (2015). Consumer buying motives and attitudes towards organic food in two emerging markets: China and Brazil. *International Marketing Review, 32*(3/4), 389–413.

Thomas, T., & Gunden, C. (2012). Investigating consumer attitudes toward food produced via three production systems: Conventional, sustainable and organic. *Journal of Food, Agriculture and Environment, 10*(2), 132–135.

Tsakiridou, E., Boutsouki, C., Zotos, Y., & Mattas, K. (2008). Attitudes and behaviour towards organic products: An exploratory study. *International Journal of Retail & Distribution Management, 36*(2), 158–175. https://doi.org/10.1108/09590550810853093

Yabe, M., Hayashi, T., & Nishimura, B. (2013). Economic analysis of consumer behavior and agricultural products based on the biodiversity conservation value. In J. R. Pillarisetti, R. Lawrey, & A. Ahmad (Eds.), *Multifunctional agriculture, ecology and food security*. Nova Science.

Wang, X., & Zhang, J. (2016). Does priming a sense of powerfulness encourage consumers to buy healthy foods? *Advances in Consumer Research, 44*, 673–674.

Welsch, H., & Kühling, J. (2009). Determinants of pro-environmental consumption: The role of reference groups and routine behavior. *Ecological Economics, 69*(1), 166–176. https://doi.org/10.1016/j.ecolecon.2009.08.009

Sustainable Water Resources Management Based on the DPSIR Framework in East and West African Countries

Pantazis Georgiou, Christos Mattas, Konstadinos Mattas, Dimitra Lazaridou, and Stefanos Nastis

1 Introduction

In recent years, there has been global concern about water availability. Water demands are increasing due to global population growth, human well-being and development, and the intensification of economic activities. Climate change projections predict that temperature will rise and heavy rainfall incidents will happen more often worldwide, impacting the environment and populations. Drought periods will be prolonged, and floods will occur more frequently. Water supply will be threatened along with food and health security.

Water covers 71% of the Earth's surface. Water shortage or pollution causes 80% of diseases that are pervasive worldwide (Siwailam et al., 2019). Approximately 97.5% of the world's water is saline, found in the oceans, and the remaining 2.5% is fresh. Glaciers constitute the largest part of freshwater (68.7%), groundwater is

P. Georgiou (✉)
School of Agriculture, Department of Hydraulics, Soil Science and Agricultural Engineering, Aristotle University of Thessaloniki, Thessaloniki, Greece
e-mail: pantaz@agro.auth.gr

C. Mattas
School of Geology, Department of Structural, Historical & Applied Geology, Aristotle University of Thessaloniki, Thessaloniki, Greece

K. Mattas · S. Nastis
School of Agriculture, Department of Agricultural Economics, Aristotle University of Thessaloniki, Thessaloniki, Greece

D. Lazaridou
School of Agriculture, Department of Agricultural Economics, Aristotle University of Thessaloniki, Thessaloniki, Greece

Department of Forestry and Natural Environment Management, Agricultural University of Athens, Karpenisi, Greece

K. Mattas et al. (eds.), *Value Chain Dynamics in a Biodiverse Environment*, Cooperative Management, https://doi.org/10.1007/978-3-031-49845-9_5

equal to 30.1%, and permafrost is 0.8%. Surface and atmospheric water are just 0.4% of total freshwater resources. Lakes contain 67.4% of surface water and rivers 1.6%. It is concluded that the total amount of freshwater that humanity can use to meet its demands is a very small percentage of the total amount of water on earth.

Assessment of environmental degradation and its impact on socioeconomic conditions has gained significant importance for the long-term management of water resources and ecosystem sustainability. Water management requires innovative practices to cover demands and provide good quality water in sufficient quantity for future generations.

Scientists and managers have created many indicators to assess the condition of water resources. Still, integrated water management requires an efficient tool to investigate the interaction between the environment and humans. The Driving Forces–Pressures–State–Impacts–Responses (DPSIR) framework is a valuable tool to investigate the above interaction.

DPSIR methodology has been widely used to facilitate empirical research, to develop indicators, and to organize the information contained in management plans. It has been utilized to investigate the causal chains of the environmental consequences of an off-shore wind farm (Elliot, 2002), to identify the social and economic pressures in an estuary (Caiero et al., 2004), for protection of groundwater, inland surface waters, estuaries, and coastal waters (Borja et al., 2006), for the assessment of impacts of development activities on the coastal environment and society (Lin et al., 2007), for determining sustainability indicators at coastal zones (Bell, 2012), to identify several environmental problems in a river basin, with the aim of designing an Integrated River Basin Management Plan (Kagalou et al., 2012), and for integrated groundwater resources management in Northern Greece (2014).

The EWA-BELT project, funded by the European Union's Horizon 2020 program, examines the human and ecological factors affecting water quality and quantity in case studies from West and East African countries (Ghana, Burkina Faso, Kenya, and Tanzania). This study aims to analyze the environmental and societal impacts, propose strategies to address the problems, and improve water resources management.

2 DPSIR Methodology

The DPSIR framework, promoted by the European Environmental Agency (European Environmental Agency—EEA, 1999), has proved to be a valuable tool for managing complex issues regarding water management (Kristensen, 2004; Carr et al., 2007; Crouzet et al., 2009). It has been applied in many cases to analyze environmental problems, specifically to identify the human and ecological factors affecting water quality and quantity, to propose strategies to address problems, and to improve water resources management.

According to Kristensen (2004), the components of the DPSIR framework are defined as follows:

- *Driving forces* include anthropogenic and natural factors.
- Driving forces lead to human activities that exert *Pressures* on the environment. There are three types of pressures: excessive use of environmental resources, changes in land use, and emissions to air, water, and soil. Natural phenomena may exert forces, as well.
- The *State* of the environment is the combination of the physical, chemical, and biological conditions.
- *Impacts* are caused by changes in the state that may have environmental or economic "impacts" on the functioning of ecosystems, on their life-supporting abilities, and ultimately on human health and society's economic and social performance.
- *Responses* proposed by societies or policy makers have to do with decision-making. They may seek to control the Pressures and affect any part of the chain between driving forces and impacts.

Their inter-relationship is depicted in Fig. 1.

A mixed approach was carried out in this study. Firstly, a search in databases was held to identify relevant works and further climate, agronomic, and socioeconomic information for the examined areas. This process contributed to better understand the current state and determine the main factors that have an influence on water resources' status. Based on the information obtained, a structured questionnaire was

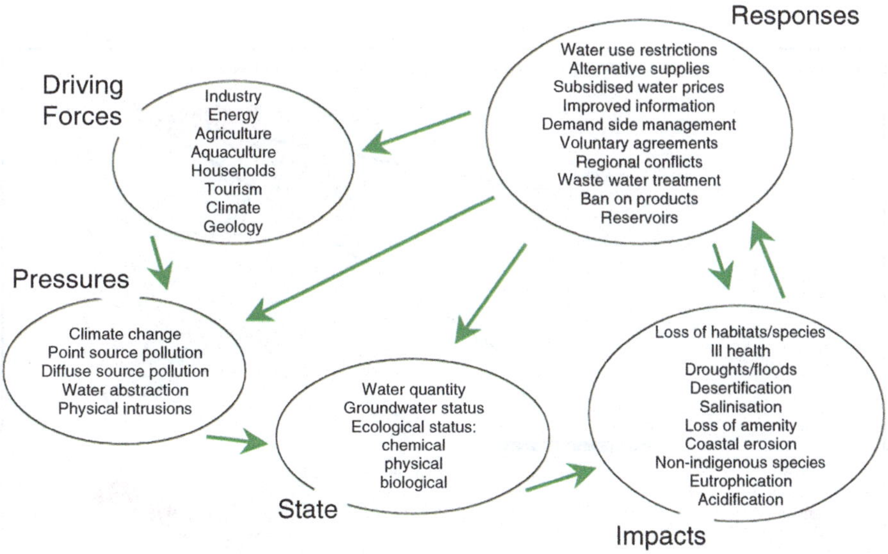

Fig. 1 A generic DPSIR framework for water (Kristensen, 2004)

designed and distributed to partners of the project in the aforementioned areas. The feedback provided by the questionnaires contributed to the establishment of the DPSIR framework, for each area.

3 Results and Discussion

3.1 Ghana Case Study

The case study area is located in the Nabdam District in the Upper East region of the country (Fig. 2). The annual precipitation in the area is equal to 950 mm. According to the Africa Groundwater Atlas (Obuobie et al., 2018), 40% of households depend on groundwater since it is considered one of the most economical and feasible sources of potable water supply despite the lack of infrastructure.

The population is primarily rural (84.3%), living in dispersed settlements. Just 15.7% of the population lives in urban areas. The Upper East region is the least urbanized area in the country. About 56% of the labor force is below 35 years. The main economic activity in the study area is smallholder agriculture (employing 80% of the population), based on rain-fed agriculture and furrows for irrigation. There is little mechanized farming. Most food crop farms are intercropped; mono-cropping is mainly associated with larger-scale commercial farms. Many people are employed in the sector of small-scale mining activities. Agricultural practices (fertilizer application) and the ineffective implementation of regulations and laws are recognized as significant drivers of water quality issues.

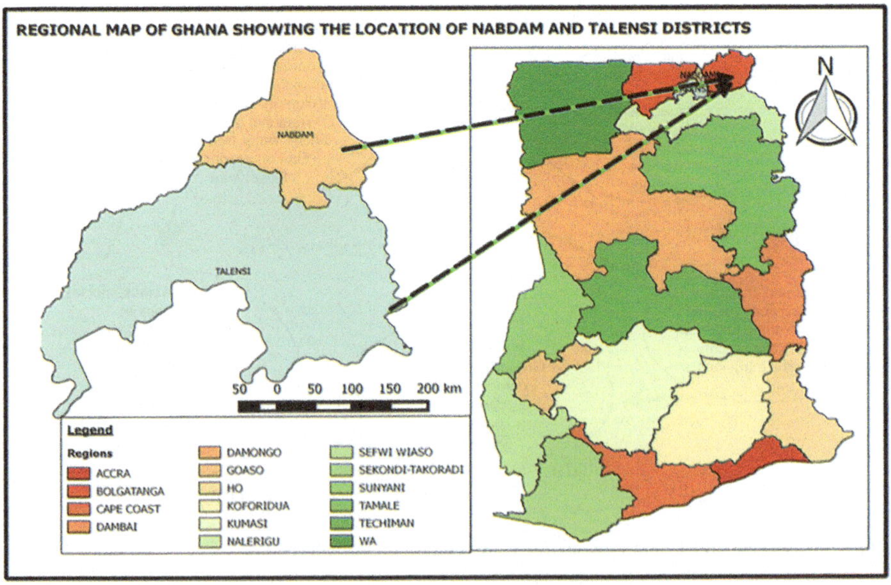

Fig. 2 Case study area in Ghana

Leveraging the gathered information, the DPSIR framework for Ghana was established. The significant factors included in the model are described below.

3.1.1 Driving Forces

The main driving forces, i.e., socioeconomic activities that exert pressures on water, can be summarized in the following: urbanization, intensive agriculture, agriculture expansion, animal breeding, industry, population growth, energy production, climate, and geology.

Agyemang et al. (2007) mention that the legalized small-scale mining activities attracted migrants into the area, leading to population growth, increasing and thus impacting the environment, including water resources. Environmental degradation is also attributed to the district assemblies' internal politics and indifferent attitudes, the regional coordinating council, and the government environmental agency. Bawakyillenuo (2020) highlights the role of illegal mining activities in water resources degradation, considering it the most destructive human activity. Agricultural practices (fertilizer application) and the lack of political will to implement regulations and laws are significant drivers of water quality issues.

3.1.2 Pressures

The main pressures that partners from Ghana identified on the local water resources are the over-exploitation of natural resources, including water used for mining activities which, in many cases, are illegal. Water quality degradation comes from bad practices accompanying mining activities (such as sand and stone mining) and mistaken farming practices that include bush burning affecting soil and water in the long term. Climate change also exerts pressure on water. Declining rainfall and higher evapotranspiration rates in neighboring Burkina Faso could reduce the Volta River's annual levels in the following years (McCartney et al., 2012).

3.1.3 State of the Water

The main outcomes of the Hydrogeological Assessment Project (HAP) of the northern regions of Ghana were published in December 2011 (Carrier et al., 2011). Data concerning groundwater quality in the broader study area are provided in Fig. 3. The pH values in most of the samples range between 6.5 and 8 (Fig. 3a), displaying the typical values of freshwater. The pH values are increased only in a small number of samples, and the water is characterized as alkaline. For iron, the suggested maximum permissible limit by the World Health Organization is equal to 0.3 mg/l for potable water, and the guideline for manganese sets the value at 0.4 mg/l. Most samples are below these threshold values (Fig. 3b, c). Almost all the samples collected from the study area (Fig. 3d) display chloride values lower than 250 mg/l, representing freshwater and within the limits set by the World Health Organization for potable water.

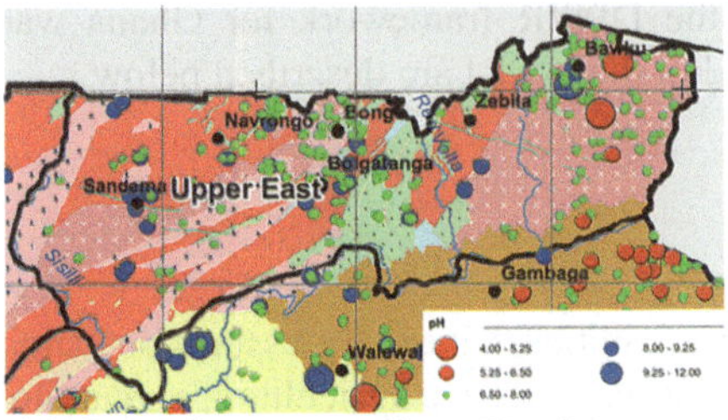

(a) Groundwater samples pH values

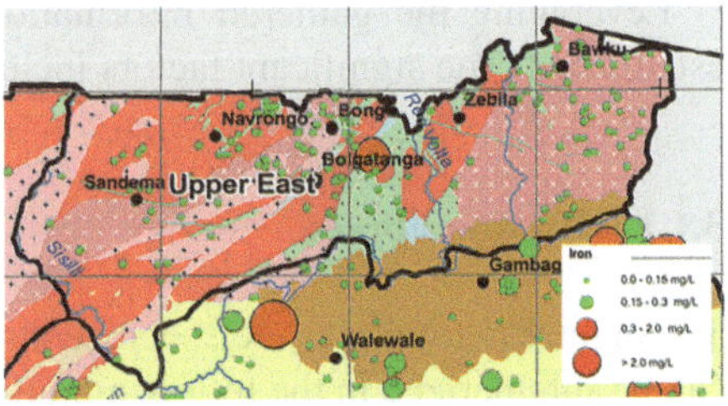

(b) Groundwater sample iron values in mg/l

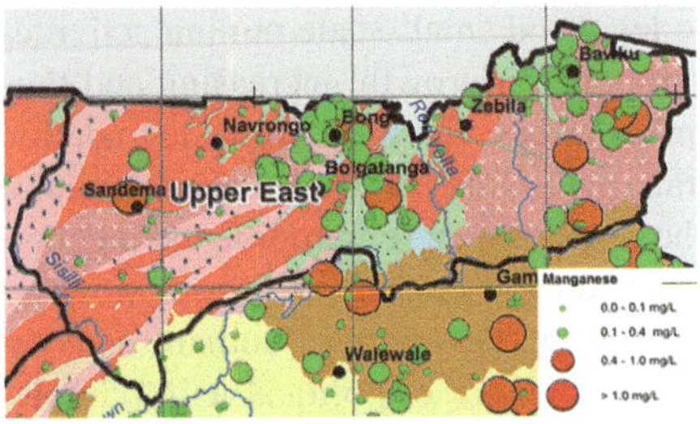

(c) Groundwater sample manganese values in mg/l.

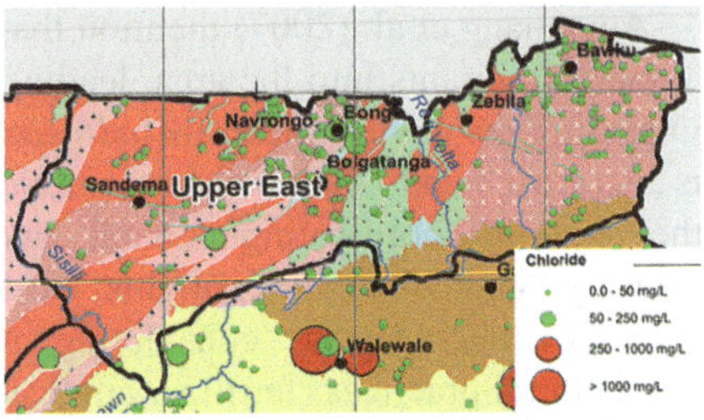

(d) Groundwater sample chloride values in mg/l

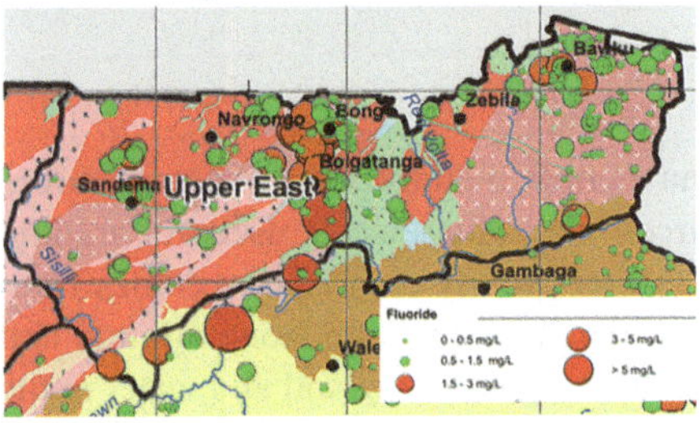

(e) Groundwater sample fluoride values in mg/l

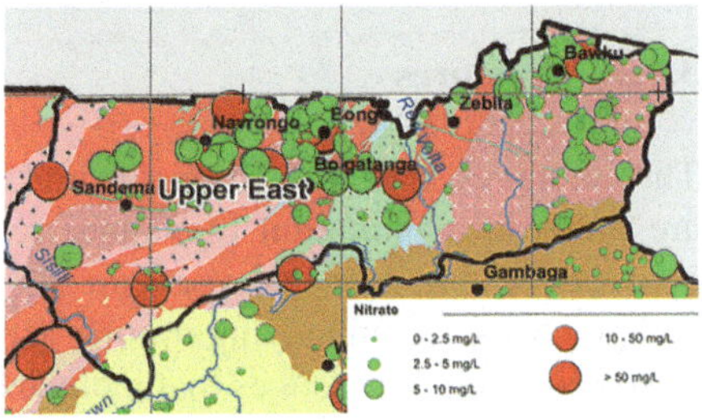

(f) Groundwater sample nitrate values in mg/l

Fig. 3 Groundwater samples values for pH, Fe, Mn, Cl, F, and NO_3. (**a**) Groundwater samples pH values. (**b**) Groundwater sample iron values in mg/l. (**c**) Groundwater sample manganese values in mg/l. (**d**) Groundwater sample chloride values in mg/l. (**e**) Groundwater sample fluoride values in mg/l. (**f**) Groundwater sample nitrate values in mg/l. Data source: Geology from Ghana Geological Survey (revised map from 2009) and all base map layers from SWERA

The samples collected from the study area show fluoride values lower than 1.5 mg/l. Many samples on the axis between Bongo and Bolgatanga show extremely high values (Fig. 3e). Nitrate-nitrogen data generally display concentrations below the 50 mg/l value set by the WHO. The highest concentrations are usually observed near large urban areas, indicating anthropogenic activities as a probable source (Fig. 3f).

3.1.4 Impacts

The major impacts on human well-being and the environment can be summarized as follows:

- *Increase of flooding and drought phenomena.* The impacts of climate change on water resources and the environment are recorded all over the country (Asumadu-Sarkodie & Owusu, 2015a, b). Flooding phenomena and drought periods happen more often. Future projections based on climate models depict an increase in the severity and frequency of extreme weather events in the near future.
- *Groundwater reserves decline.* According to recent measurements, the groundwater withdrawals are lower than the renewable supply. However, groundwater level draw-down (implying reserves reduction) is recorded in many locations since many wells are clustered and unregulated. Groundwater availability and risks stem from poor regulation and enforcement, a lack of reliable data, drought, and reduced recharge from climate change.
- *Reduction of hydropower generation, inland aquaculture, and agriculture across northern Ghana.* Over the last few decades, drought has caused national hydroelectric production to fall to 33% capacity in some cases. Higher temperatures and drought will reduce agricultural production in the country's northern areas.
- *Population displacement.* Between 1988 and 2010, Ghana experienced 15 significant floods that displaced hundreds of thousands of people. Climate change has increased the flood risk of urban areas and riverine communities.
- *Health issues.* The ratio of boreholes per capita in the broader study area needs to be improved since it cannot provide good quality groundwater to the communities, and they are forced to use alternative water sources.

3.1.5 Responses

- *Incorporation of customary water management.* Before the colonial era, traditional leaders used to manage water resources successfully. This was achieved using empirical methods such as plowing, contouring, water storage in clay pots, digging wells, homestead ponds, and rainwater harvesting.
- *Strengthening of collaboration among stakeholders-authorities and assignment of specific duties.* The human activity with the most significant negative effect on the environment, specifically in Ghana's water bodies, is small-scale illegal mining. This industry type has become a major employer to many people in rural areas.
- *The strong political will to strengthen and implement existing water-related policies.* There are existing comprehensive water-related policies, laws, and regulations that are not adequately enforced and, therefore, ineffective in addressing the problem (Yeleliere et al., 2018). Actions such as banning specific plastic materials from the market, surveillance, imposition of fines, and implementing pricing policies could effectively sustain water resources for the next generations.
- *Education and sensitization of citizens.* Solid and long-term collaboration between the media, education system, and competent authorities (such as the Ministry of Information, local governments, and the National Commission for Civic Education) is required to increase awareness of citizens on water pollution threats and the risk it poses for the present and the future.

- *Investment in wastewater treatment and waste re-use*. Sorting waste into components such as plastics, glass, metals, and organic at the household level is crucial for sustainable waste management in Ghana. Along with good rubbish collection, disposal and proper landfill management practices will support the effort for environmental and, therefore, water sustainability.
- *Renewable energy resources for mitigating climate change*. Climate change has become a global phenomenon. The use of alternative energy sources (such as liquefied petroleum gas, which has been promoted by the Government of Ghana under specific actions in recent years) and usage of renewable sources such as solar, wind, and hydro power could effectively reduce Ghana's footprint on emissions (Asumadu-Sarkodie & Owusu, 2016a, 2016b, 2016c).
- *Construction of facilities*. The number of boreholes constructed must be increased. Well-planned reforestation and rainwater harvesting schemes should be implemented to increase groundwater quantity. It is also of major importance that stakeholders using the existing facilities contribute to the maintenance and proper use of the equipment, which is usually a hand pump.

The DPSIR framework for the area is presented in Fig. 4.

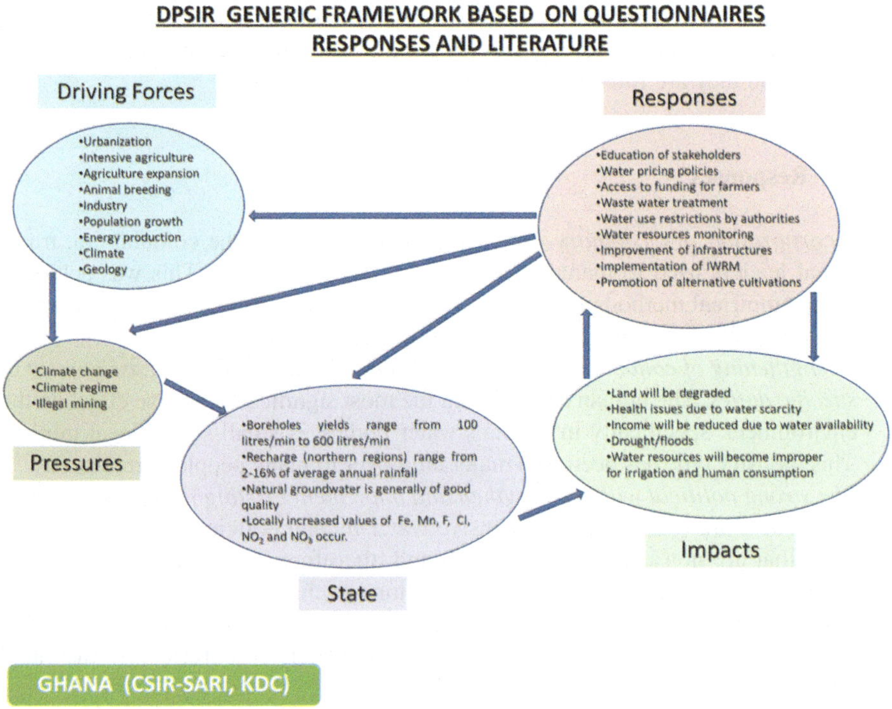

Fig. 4 DPSIR framework for water resources in the Ghana case study

3.2 Tanzania Case Study

The case study area is located in the Arusha Region in the Upper East Region of the country (Fig. 5). Tourism is one of the main economic activities in the Arusha Region, along with agriculture. The majority of the population in large urban centers depends on groundwater to meet their demands.

Many foreign horticultural agribusinesses and agro-industries operate in the area. Smallholder farmers also practice agriculture (Komakech & de Bont, 2018). The main cultivated crops are coffee, grain, vegetables, cotton, pyrethrum, papain, sisal, and sunflower seeds. Mining activities take place in the area. Tourism is one of the main economic activities in the Arusha Region, along with agriculture. Due to the favorable climatic conditions, many foreign horticultural agribusinesses and agro-industries operate there.

Fig. 5 Selected area for the implementation of the DPSIR framework in Tanzania

After a literature review and feedback from questionnaires disseminated to partners from Tanzania, the DPSIR framework was established.

3.2.1 Driving Forces

The main driving forces, i.e., socioeconomic activities that exert pressures on the water, are: population growth, geology, intensive agriculture, agriculture expansion, animal breeding, climate, and energy production.

The population of Tanzania has tripled from 1968 to 2012. The region's population doubled over the years 1988 and 2012. The annual population change is 2.8% for 2002–2012 (https://www.citypopulation.de/en/tanzania/admin/02__arusha/, accessed on 1 December 2021).

Therefore, the water demands are increasing along with the threats (pollution sources) for water resources. Groundwater quality can be affected by anthropogenic pollutants, as well. According to Ghiglieri et al. (2012), the interaction between the groundwater and fluoride-rich minerals and other natural processes could explain the observed variation in fluoride concentration in the Arusha Region.

The country has one of the largest livestock populations in East Africa (CIAT & World Bank, 2017). The most common livestock types are traditional breeds of cattle, sheep, goats, poultry, and pigs (FEWS NET, 2018). Livestock is mainly produced in extensive systems practiced by pastoralists and agro-pastoralists on natural pastures. Intensive and semi-intensive systems are standard for improved livestock breeds.

3.2.2 Pressures

The main pressure regarding Tanzania as identified in the local water resources is the uneven rainfall distribution in space and time. Due to the recorded population growth in the Arusha Region, a significant amount of land has changed use and was converted to settlements or agricultural land. This has a negative impact on groundwater quality and quantity. Water demands for domestic purposes and food production have increased, resulting in increased water abstraction. Aquifers' recharge rate is reduced since buildings, road networks, and other types of infrastructures occupy natural land. Water pollution from agriculture due to agrochemicals and fertilizers and from untreated domestic and industrial waste has increased. In combination with the operation of mining industries, the growth of tourism has increased Arusha city's water demand (Komakech & van der Zaag, 2011).

3.2.3 State of the Water

A detailed study concerning groundwater quality and quantity in the Arusha District (bordered by the three administrative districts of Monduli, Longido, and Meru) is provided by Chacha et al. (2018). The main results of this research are the following: (1) The quality of groundwater is suitable for drinking purposes and most of the

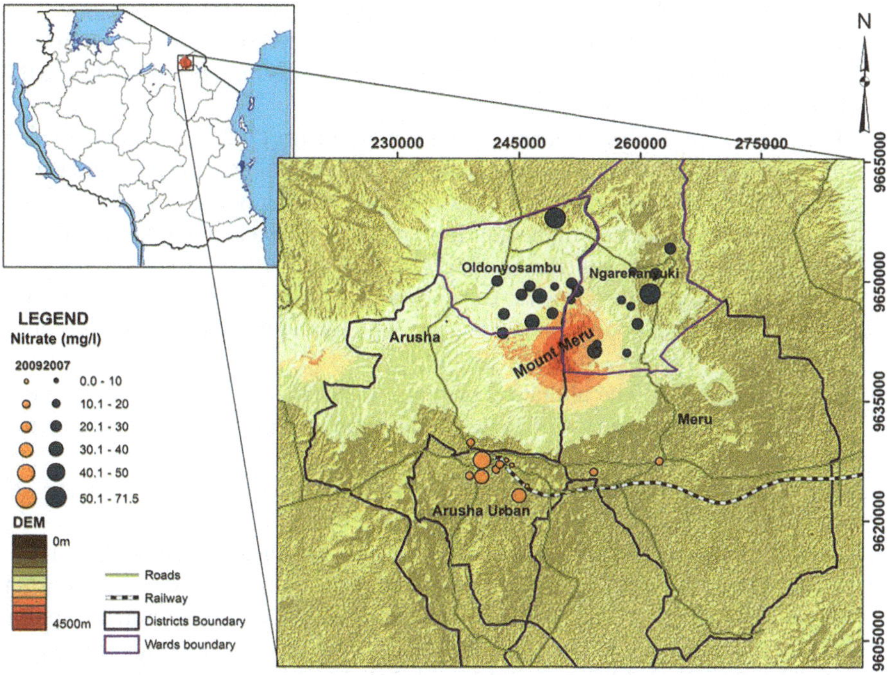

Fig. 6 Spatial distribution of nitrates concentration in groundwater samples for the Arusha Region (Elisante & Muzuka, 2017; GITEC & WEMA, 2011; Pittalis, 2010)

other common uses, except fluoride content. (2) Quality in the study area is dominated by sodium and bicarbonate ions, which define the general composition of the water type to be Na–K–HCO_3. (3) The groundwater chemistry is mainly determined by aquifer lithology and less by anthropogenic activities. (4) The most important source of anthropogenic pollution is the agricultural activity that takes place in the area, as it is confirmed by the presence of nitrates in the groundwater, mainly in shallow wells and springs (Fig. 6). (5) The depth from the ground surface to the groundwater level ranged from 2.7 to 93.2 m for public wells and 1.98 to 7.03 m for privately owned wells drilled in the area.

3.2.4 Impacts

The main impacts on human well-being and the environment can be summarized as follows:

- *Increase of drought and flooding phenomena.* According to the climate models' projections, precipitation in Tanzania will increase over the next few years along with the temperature. The most worrying issue is that extreme drought probability tends to increase by 4–13%, due to changes in the inter-annual and inter-seasonal precipitation patterns.

- *Water supply and food shortages.* The prolonged drought periods that have already occurred in the past (such as in 2003 and 2005) resulted in significant livestock loss and damaged farmers. Food security was threatened (FCFA, 2017; Oikos, 2011).
- *Decrease of groundwater reserves.* Olarinoye et al. (2020) ran mathematical simulation models to investigate the effect of climate change and the rapid urbanization of Arusha city on the groundwater. The results revealed that there would be a significant reduction in the aquifers' recharge rate and a major groundwater level draw-down of many tens of meters.
- *Income reduction of the farmers.* Dependence of income on rainfed farming constitutes small-scale poor farmers' income which is vulnerable to droughts since they affect crop yields (Mdemu, 2021).
- *Energy production.* The alteration of rainfall and, therefore, the flow regime of the hydrographic network will have an impact on water use for power production (Molina et al., 2020).
- *Health issues.* Flooding phenomena usually favor the spread of waterborne and insect-borne diseases (Paavola, 2004).
- *Land degradation.* In the frame of the Land Degradation Neutrality Target Setting Programme (LDN TSP), a National Working Group (NWG) was established. The main drivers identified for the entire territory of the United Republic of Tanzania were the following: overgrazing, poverty, the land tenure system, scarcity of firewood/charcoal making, population increase, poor farming practices, and climate change.

3.2.5 Responses

As has been mentioned, water availability in the future will be aggravated under climate change, and water scarcity problems will be added to the existing ones, since prolonged drought periods will occur more often. Participation of local stakeholders and end-users should be increased in the decision-making process concerning water management issues. A long-term solution in the basin that will improve irrigation efficiency and reduce water losses is the conversion of furrows to a pipeline network. Better farming practices, such as contour stone bunding, tied contour ridges, terrace farming, and others, should be promoted and applied to a larger extent (Shaghude, 2006). Artificial recharge of aquifers could also be a very efficient solution in environmental, social, and economic terms.

Partners from Tanzania, based on their knowledge, experience, and expertise, highlighted the following responses as the most effective to achieve water resources sustainability in the area: Education of stakeholders, water pricing policies, access to funding for farmers, wastewater treatment, water resources monitoring, improvement of infrastructures, implementation of IWRM, and promotion of alternative cultivations.

The produced DPSIR framework is depicted in Fig. 7.

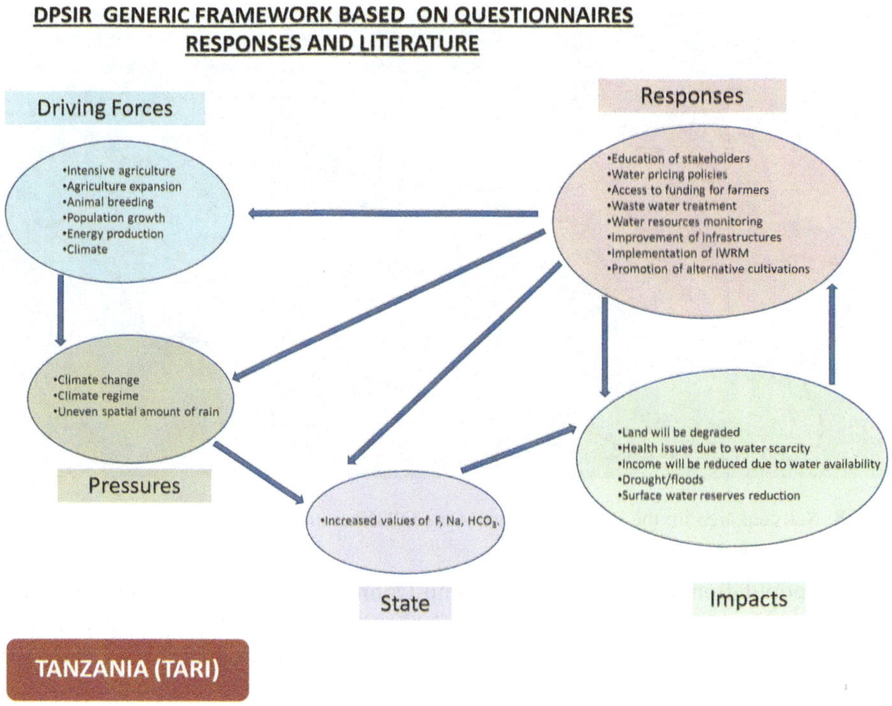

Fig. 7 DPSIR framework for water resources in the Tanzania case study

3.3 *Kenya Case Study*

The case study area is located in the Nyakach Region in the west part of the country (Fig. 8). The Nyakach Region has low water coverage, poor water management, and inadequate sanitation. As a result, health issues related to diarrheal infections, such as cholera and a high incidence of diarrhea, have been recorded in the area. Most inhabitants are occupied in agriculture, fishing, and sand harvesting (Wasonga et al., 2016). The main crops cultivated include maize, sorghum, beans, sugarcane, rice, and cotton. Small-scale production represents 90% of total agricultural production and 70% of agricultural production available in the county's market (MoALF, 2017). Floods, droughts, and heat stress have become more frequent and intense, impacting the crop's productivity and, therefore, people's income. Agriculture in the area is mainly rain-fed, but large parts of agricultural land are irrigated by surface or groundwater. Industries, especially some agri-food factories, abstract water, as well.

3.3.1 Driving Forces

The main driving forces, i.e., socioeconomic activities or natural processes that exert pressures on water, are population growth, urbanization, intensive agriculture, agriculture expansion, industry, and climate.

Fig. 8 Selected area for the implementation of the DPSIR framework in Kenya

The population growth rate for Kisumu County was 1.8% for 2009–2019, and the urbanization rate was equal to 38% (1980–2020) (https://www.citypopulation.de/en/kenya/admin/, last accessed February 24, 2022). The evolution of the Nyakach sub-county population is depicted in Table 1. The Nyakach sub-county is a very densely populated area (Table 1) compared to the national level, equal to approximately 92 persons/km^2 (https://www.worldometers.info/demographics/kenya-demographics/). The majority of the population (68.8%) resides in rural areas. The mean household water consumption is 149.50 l/day with a mean per capita of 32.92 l/day for the entire Kisumu County (Wagah et al., 2010).

Water demands for human consumption and irrigation demands will be significantly increased, threatening food security and drinking water supply. The increased demands for food production will result in intensifying agricultural activities and probable expansion of agricultural land. Irrigation water will be of utmost importance, and its management will be essential for inhabitants. According to Owen (2020), the existing irrigation systems in Kisumu County need rehabilitation of deteriorated facilities due to insufficient maintenance. Therefore, irrigation is already considered a driver that negatively affects water resources. If future scenarios on irrigation demands are met, then agriculture will become one of the most important driving forces exerting pressure on the water.

3.3.2 Pressures

The main pressures that partners from Kenya identified on the water resources are climate change and illegal mining activities. The climate change phenomenon intensifies in Kenya as the country is already experiencing high temperatures over the last few decades (Herrero et al., 2010). Many rivers' discharge rates and water levels will decline,

Table 1 The population of the Nyakach sub-county over the years (KISUMU COUNTY INTEGRATED DEVELOPMENT PLAN II, 2018–2022)

Year	2009	2018	2020	2022
Population	133,041	168,140	177,128	186,582
Density (persons/km^2)	372	470	495	522

affecting electrical energy production (Bunyasi, 2012). Both surface and groundwater abstractions will be significantly increased to meet the increased demands under climate change and population growth. Point and diffuse pollution is estimated to be expanded, following the ongoing growth of urban areas due to the lack of sewage networks, an existing problem for many years (Kanoti et al., 2019; AFW, 2014).

3.3.3 State of the Water

Ajuang et al. (2016) conducted a questionnaire survey and interviews in the Upper Nyakach area to collect qualitative and quantitative data. Their study sought to determine household climate change awareness levels using common climate change indicators such as heavy rainfall incidents, flooding events, drought periods, and temperature. According to the findings, the majority of the households (86.7%) reported that they had observed changes in water sources. The actual changes that were observed were the following: drying up of water sources (63.3%), reduction in water quality (17.3%), conflicts over water access (12.7%), increasing distance to water sources (5.8%), and a rise in the prevalence of waterborne diseases (0.9%).

According to Oiro (2018), groundwater flow follows the direction of the hydrographic network. It flows from the high-altitude areas toward Lake Victoria. Groundwater accounts for 15% of all irrigation withdrawals (mainly in the Rift Valley Basin). There is potential for further groundwater exploitation; therefore, the National Water Master Plan projects that 8% of new irrigation abstractions by 2030 will depend on groundwater (USAID). Boreholes in the Lake Victoria Basin are deep in most locations. The average depth ranges between 50 and 150 meters (m). Shallow aquifers are developed in the coastal plains near Lake Victoria and alluvial fans along with river flow (Barry & Obuobie, 2012).

3.3.4 Impacts

The impacts of climate change in Kenya have affected some of the key sectors of the country's economy (GoK, 2007; Kuria, 2009; SEI, 2009). Kenya is naturally prone to drought due to high inter-seasonal and inter-annual variability. Severe droughts were recorded in 2010–2011, 2016–2017, and 2019 (FAO, 2020; Uhe et al., 2016). The latter prolonged drought period affected more than 3 million people. 25% experienced malnutrition since rain-fed agriculture is the prevailing cultivation pattern, and 60% of livestock are in arid and semi-arid areas (Walsh, 2019). Increased numbers of extreme rainfall events, combined with land-use changes (deforestation for agricultural expansion), cause floods that intensify soil erosion and result in siltation of watercourses, threaten infrastructure and human lives, spread water-borne diseases, and cause economic losses due to the damages caused to crops (NEMA, 2011). The entire Kisumu County is an extremely high-risk area for both drought and flooding phenomena (Fig. 9a, b).

The rapid population growth, the growing commercial activities, industrialization, and insufficient waste/water management in Kisumu and Lake Victoria Basin have increased the amount of urban waste. The lack of appropriate treatment facilities and technologies has degraded water resources quality (Juma et al., 2014). Some years ago, Scheren (2003) investigated the impacts of water pollution in the Lake Victoria catchment by applying the DPSIR framework. According to the findings of his

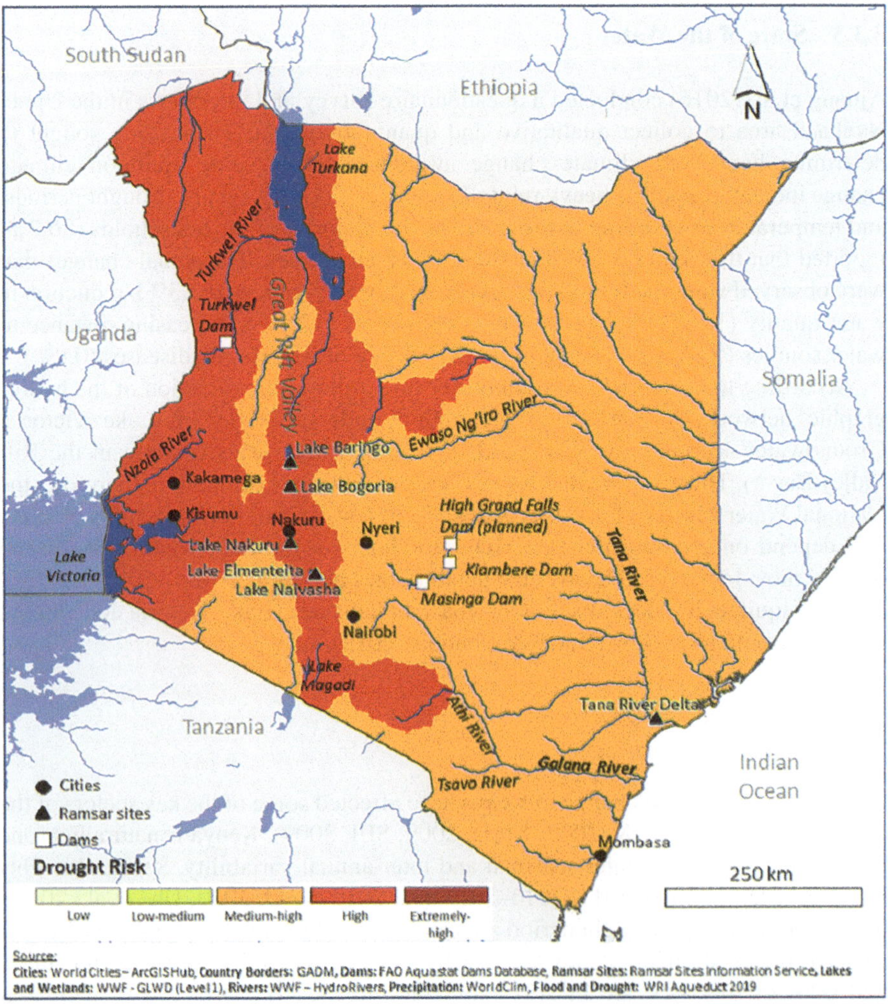

Fig. 9 (**a**) Drought risk of Kenya; (**b**) Flood risk of Kenya

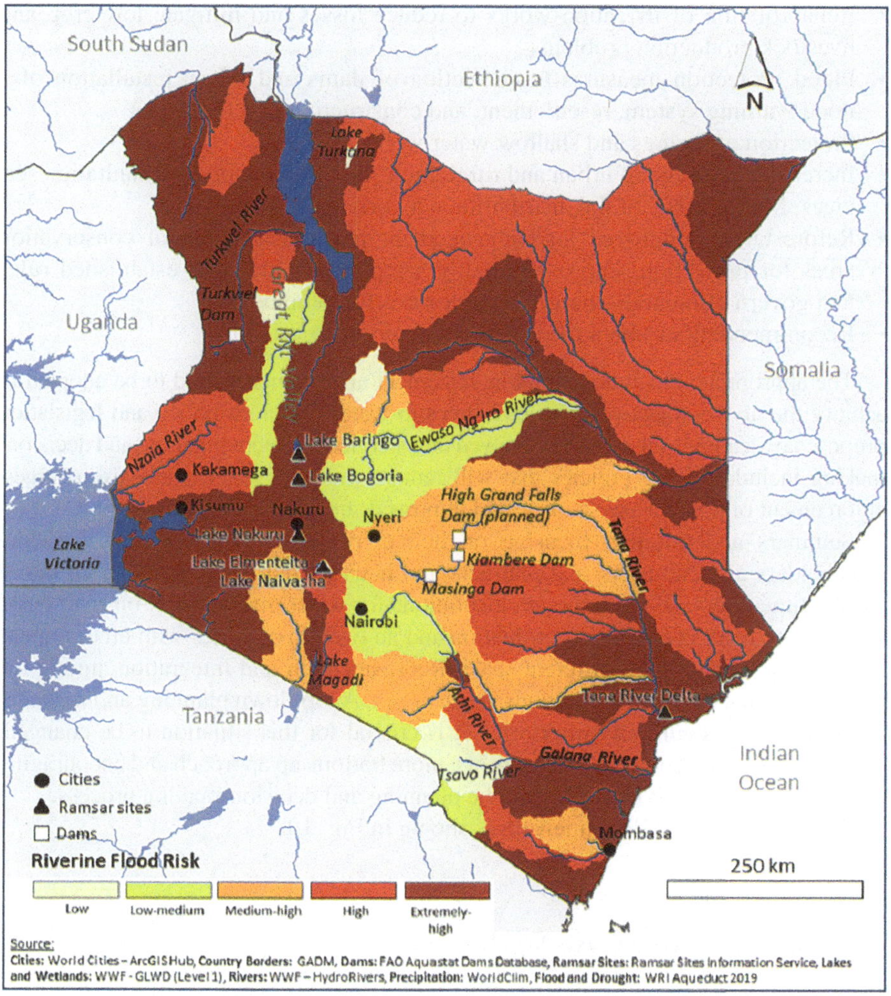

Fig. 9 (continued)

publication, it highlighted the following impacts: decrease of traditional fish resources, loss of amenity (stench and visual impacts), loss of recreational and tourism value, health issues (waterborne diseases: diarrhea, bilharzia, typhoid, dysentery, and cholera), loss of water provision value, and loss of biodiversity value.

3.3.5 Responses

The Nyando River discharges into Nyakach Bay in Lake Victoria. The basin includes parts of Nyakach County. Njogu (2000) provided a detailed list of remedial measures focusing on the conservation and protection of the area's water resources based on the findings of his research, which are presented below:

- Rehabilitation of irrigation works to reduce losses and mitigate low crop and livestock production problems
- Flood protection measures (construction of dams and dikes, installation of a flood warning system, resettlement, and construction of a reservoir)
- Protection of springs and shallow water wells
- Increased coverage of urban and rural domestic water supply and sanitation services. Maintenance of the distribution network to reduce losses
- Reforestation, improved land management practices, zoning of conservation areas for protection, improved quality sampling frequencies, established rules that govern effluent discharges by industries/households, etc.
- Recommendations about the use of fertilizers and pesticides

The appropriate institutional and management arrangements need to be adopted to achieve the above objectives. Some changes to water resources policy and legislation proposed by leading stakeholders involved in water management planning and decision-making include pricing policies that will reduce waste of water, water use charges, enforcement of polluter-pays policy, and improved surveillance and monitoring.

Seminars and training, focusing on human resources development, including stakeholders and end-users (e.g., farmers and households), should be carried out at regular time intervals. Even though institutions for water management on river basin scales exist, there is a gap in coordination and an overlap of duties with other organizations that operate on a different scale. Reorganization and integration among the functions of the different institutions are required. A top-down planning and management approach is currently in priority. It is crucial for this situation to be changed, involving local stakeholders and users in a more bottom-up approach and encouraging all interested parties to participate in the planning and decision-making process.

The produced DPSIR framework is shown in Fig. 10.

3.4 Burkina Faso Case Study

The case study area is located in the Bereba District in the Province of Tuy in the Region of Hauts-Bassins in the west part of the country (Fig. 11). The majority of the population in the area is occupied in agriculture. The main crop is cotton in rotation, with cereals irrigated by rainwater. In 2014, 49% of people living in urban areas and 81% in rural areas (in Hauts-Bassins) lived with a cost below $3.10/day. The broader area is characterized by two main seasons regarding its precipitation regime. The rainy season corresponds to cropping periods (May–October) and the dry from November to April. The average temperature in Béréba is 28.4 °C, and the average annual precipitation is 758.5 mm/year. According to the Köppen-Geiger classification, the area is classified as desert climate (BWh). August is the wettest month with an average precipitation equal to 186.17 mm, and April is the warmest month showing a mean temperature of 37.1 °C (https://tcktcktck.org/burkina-faso/hauts-bassins/bereba, last accessed 01/02/2022). Burkina Faso ranks 138th out of 169 countries in water vulnerability to climate change (ND-GAIN, 2021). Over the last few decades, extreme weather events, such as prolonged drought periods and

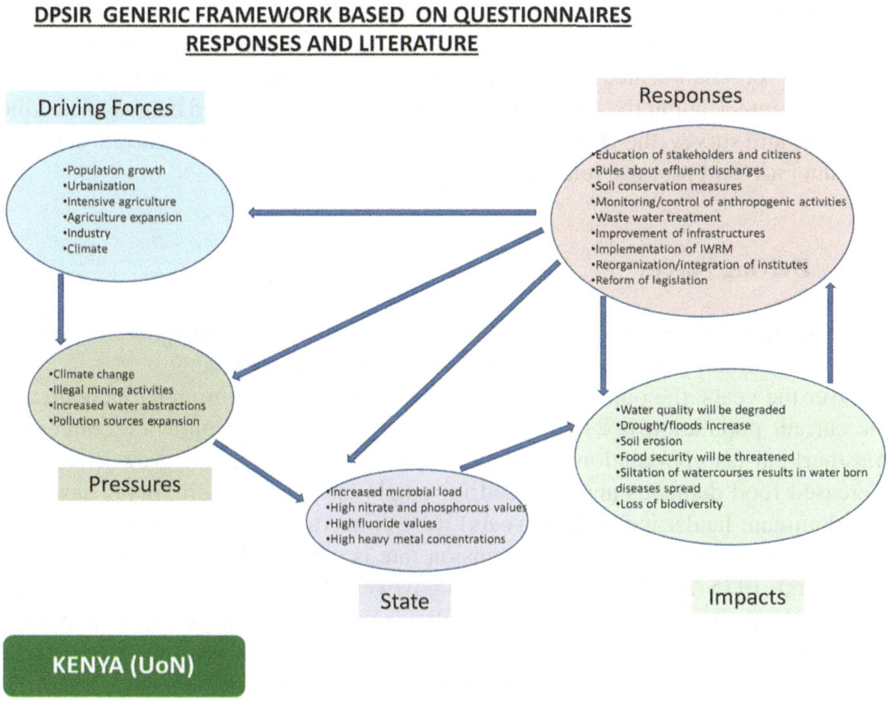

Fig. 10 DPSIR framework for water resources in the Kenya case study

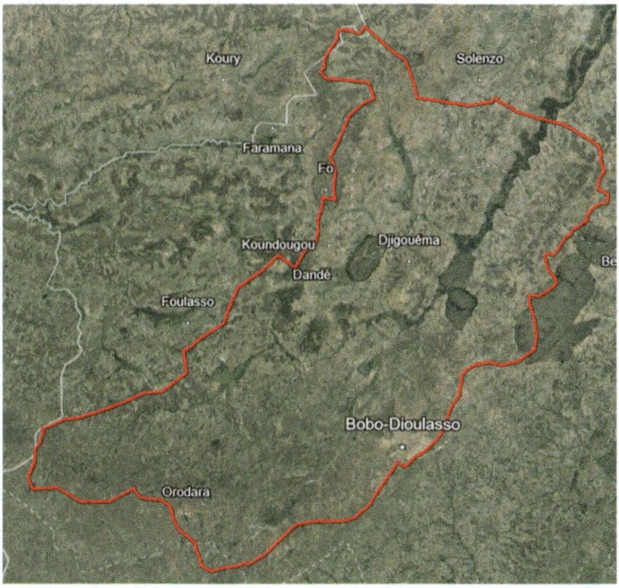

Fig. 11 Selected area for the implementation of the DPSIR framework in Burkina Faso

flooding phenomena, are recorded more often (Crawford et al., 2016). Groundwater is used mainly for drinking water supply, whereas a small percentage of groundwater is used for small-scale market garden irrigation and livestock watering.

Taking into account the information gathered from the selected databases and the questionnaire survey, the DPSIR model was structured to assess the current status in Burkina Faso and propose relevant management responses.

3.4.1 Driving Forces

The main driving forces, i.e., socioeconomic activities that exert pressures on the water, are: animal breeding, population growth, and climate change.

Over the years, the population of Burkina Faso shows a continuing increase and the current population is 21.85 million (https://worldpopulationreview.com/countries/burkina-faso-population, accessed 02/02/2022). Population growth leads to increased food demands and the need for employment. Therefore, many savannas (the dominant landscape in those years) were converted into croplands to meet the ongoing needs. The agricultural expansion rate is estimated to be 4% per year for the period 1975–2013 (https://eros.usgs.gov/westafrica/land-cover/land-use-land-cover-and-trends-burkina-faso). The population of Tuy Province showed an average annual growth rate equal to 2.9% (2006–2019). At the same time, the population of Bereba is enhanced as well, showing an increased rate of 1.7% for the corresponding years. The population density in Tuy Province is equal to 58.43 people/km^2.

3.4.2 Pressures

The variation of the climate conditions due to climate change is highlighted by the stakeholders as the main pressure on the water resources in Burkina Faso. Furthermore, the groundwater level is decreased, meaning that available water reserves are shrinking. Increased local water abstraction constitutes another significant pressure resulting from population growth, urbanization, intensive uncontrolled mining activities, agriculture expansion, and population displacement due to armed conflicts in the country after 2016 (https://www.icrc.org/en/document/burkina-faso-water-scarcity-conflict, accessed 02/02/2022; Sanon et al., 2020). Pollution of water resources from gold mining and agriculture has been an additional pressure over the last few years.

3.4.3 State of the Water

A detailed study concerning groundwater quality and quantity in Burkina Faso is provided by Martin and van de Giesen (2005). The main results of this research work are the following:

1. Groundwater production through boreholes, modern hand-dug wells, and piped systems has increased substantially over the past few decades in the

Volta River Basin. It has made groundwater an essential water source for rural and urban water supply. The groundwater consumption per person in the broader study area is depicted in Fig. 12 and is classified as one of the highest in the country.

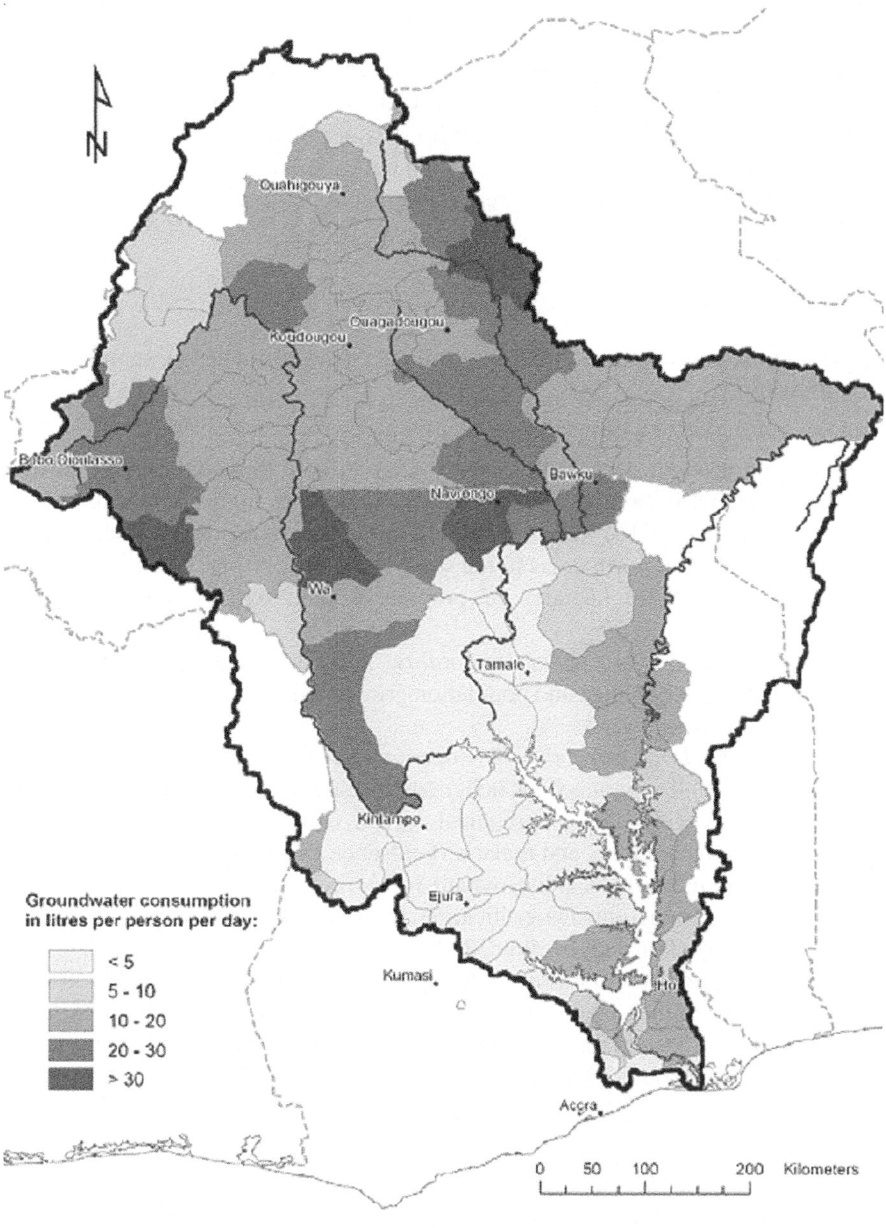

Fig. 12 Average daily per person groundwater consumption (Martin & van de Giesen, 2005)

2. The ratio of groundwater production to aquifer recharge in parts of the study area exceeds the average national ratio (Fig. 13); it is indicative, though, that there is potential for further exploitation of groundwater without posing any threats to the aquifers or having a significant effect on the hydrological balance of the area.

Although the groundwater quality in Burkina Faso is considered good to meet people's demands (Groen et al., 1988), pollution problems occur locally, especially in shallow aquifers (Yameogo & Savadogo, 2002). The use of fertilizers and chemicals in agricultural activities and the uncontrolled disposal of untreated waste from industries constitute the most significant sources of pollutants in groundwater. Due to poor sanitation facilities and services, high nitrate values in groundwater samples are recorded in densely populated areas (Groen et al., 1988).

3.4.4 Impacts

The main impacts on human well-being and the environment can be summarized in the following:

- *Land degradation.* The increase in the frequency of extreme rainfall events and prolonged drought periods potentially leads to soil degradation and erosion. Therefore, fertile land is at risk of being lost. The high deforestation rates (Crawford et al., 2016) will contribute significantly to this loss. Desertification is expanding from north areas to the south (GFDRR, 2011). It is estimated that an area of 360,000 hectares of productive land will be lost every year (FAO, 2021).
- *Decrease of per capita water availability.* The Potsdam Institute (2020) estimates that climate change and population growth will reduce per capita available water in the next 60 years.
- *Increase of health vulnerability.* The upcoming changes in the climate, temperature, and rainfall regime affect infectious disease rates. Climate change shifts timing, seasonality, and geographical spread of disease epidemics (USAID, 2012) such as meningitis and malaria (Feldscher, 2018; USAID, 2012).
- *Undernutrition.* Water scarcity, which is aggravated by climate change, may reduce food production. Desertification decrease of crop yields will increase food insecurity (Potsdam Institute, 2020), as well.
- *Conflicts and people displacement.* Desertification and decreased access to water sources have amplified tensions and conflicts between different groups of people (e.g., pastoralists and farmers) in recent years. People compete to achieve access to critical water sources and fertile land (Relief Web, 2020).

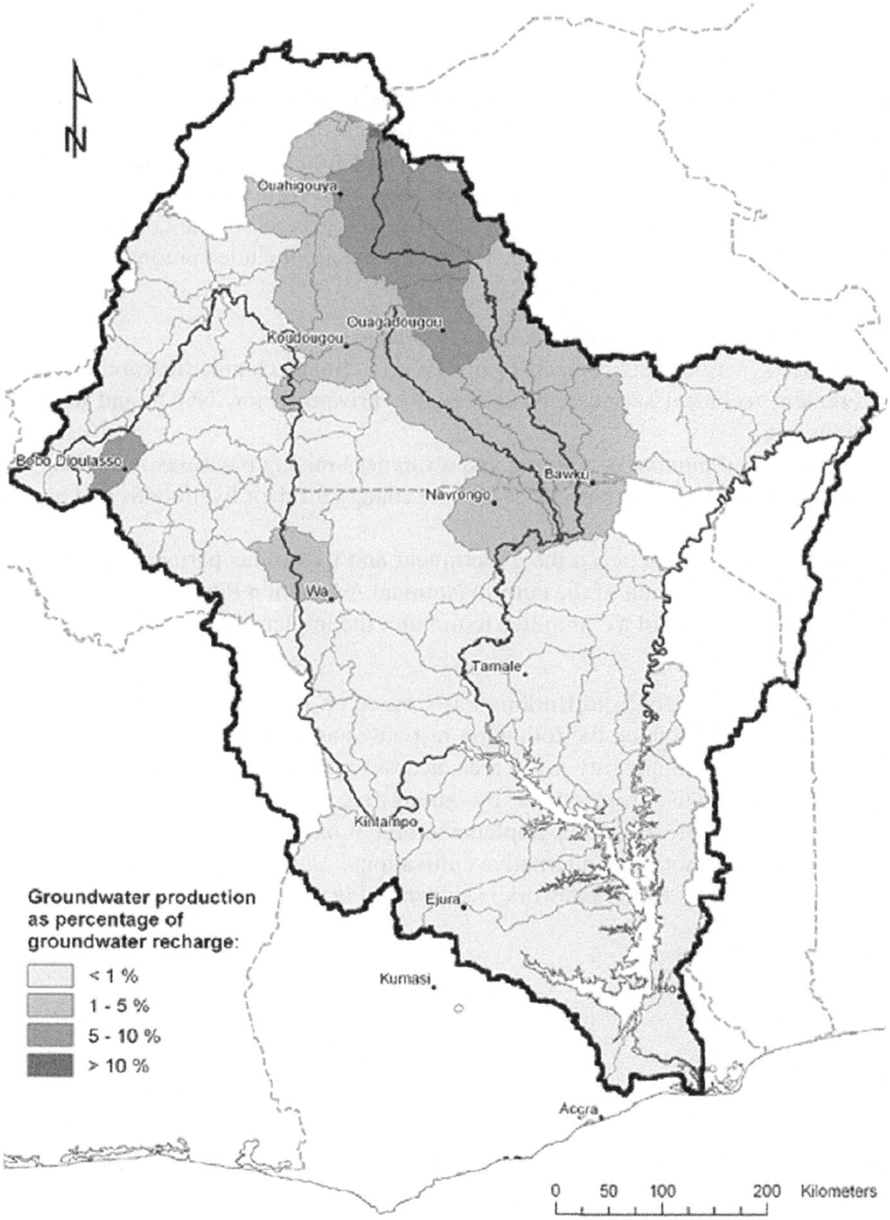

Fig. 13 The ratio of groundwater production to groundwater recharge (Martin & van de Giesen, 2005)

3.4.5 Responses

Burkina Faso has two main export products, which are cotton and gold mining (CIA, 2021). The first one is a water-intensive and heat-sensitive cultivation. It is significant for farmers to adopt alternative crops that are resilient to climate change with less water demand. In the National Climate Change Adaptation Plan (Ministry of Environment and Fishery Resources-Burkina Faso, 2015), specific long- and short-term measures for preserving and protecting the country's water and natural resources are described. The following are included among the actions that are proposed to be adopted:

- A national teaching strategy on climate change
- Greater synergies between parties involved in climate change (researchers, government technical services, producers, the private sector, NGOs and associations, etc.)
- An increased number of training and awareness-raising measures informing the public about the consequences of climate change for their livelihoods and means of subsistence
- Regular meetings between the government and its various partners with a view to the implementation of the current National Adaptation Plan
- The establishment of a committee to monitor the implementation of the National Adaptation Plan

Finally, the partners from Burkina Faso, based on their knowledge, experience, and expertise, highlighted the following responses as the most effective to achieve water resources sustainability in the area: access to funding for farmers, wastewater treatment, water use restrictions by authorities, water resources monitoring, improvement of infrastructures, implementation of integrated water resources management, and promotion of alternative cultivations.

The produced DPSIR framework is shown in Fig. 14.

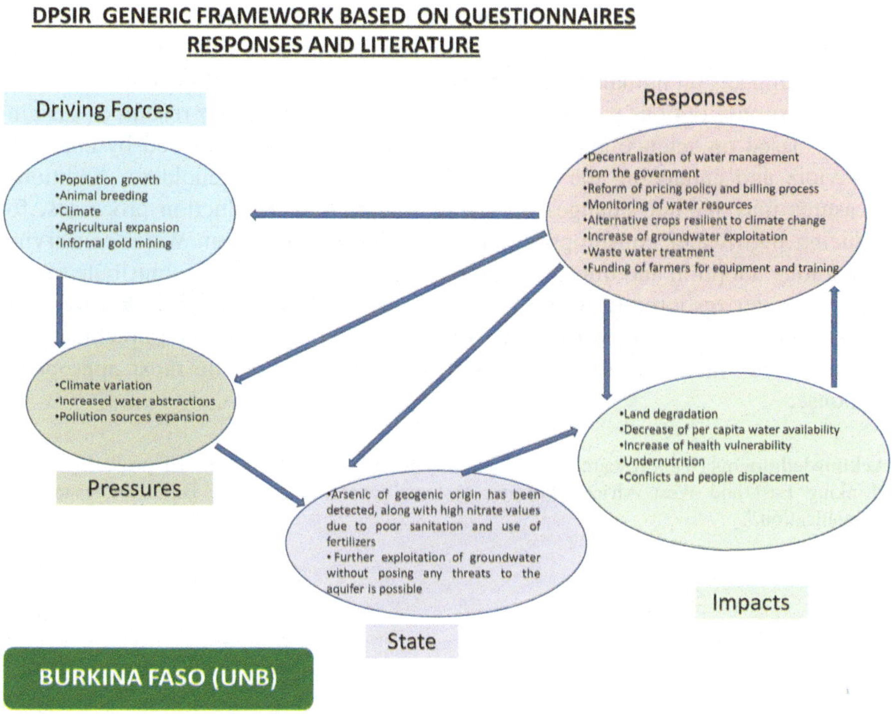

Fig. 14 DPSIR framework for water resources in the Burkina Faso case study

4 Conclusions

One of the main challenges of water management in the foreseeable future is to integrate socioeconomic parameters with natural processes. The development of societies and economies and their relationship with water resources systems require holistic management unlike the fragmented approaches implemented in the past. The DPSIR model is regarded as an appropriate tool for such integration. In this chapter, it is used to organize information about the factors affecting water quality and quantity in West and East African countries (Ghana, Burkina Faso, Kenya, and Tanzania), focusing on those that exacerbate the water budget deficit under the EWA-BELT project. DPSIR will allow decision-makers and stakeholders to view the water sustainability issues more comprehensively. Therefore, the project impact will have a long lifespan, since the implementation of the suggested measures will accelerate the achievement of water sustainability through appropriate water management. Preliminary results indicate that there is an overall increase in the driver of water consumption, in the selected case studies. Changes of climate, together with the population growth, can generate conditions of water scarcity and stress. The intensive agriculture production has put high pressure on local water resources, as well. In addition, lack of infrastructure and irrational management practices aggravate the unsustainable utilization of water resources. Further findings indicate that farmers' income is decreasing and health issues arise over the years, generating negative impacts on human well-being and on sustainable development.

The response options are measures to improve regional water resources sustainability, based on science and on serving society. It can be achieved by changing behaviors and enhancing the education and funding of stakeholders. Additional measures may include changes during the agricultural production processes, for reducing high water consumption and high nitrogen pollution. Water-conserving technology adoption and improvement of infrastructures can also contribute to sustainable water resource utilization, through the establishment of monitoring networks and the increase of wastewater treatment usage. Overall, the implementation of integrated water resources management is identified as the most appropriate response.

Acknowledgments This research was funded by the EU H2020 EWA-BELT project [862848] "Linking East and West African farming systems' experience into a BELT of sustainable intensification."

References

AFW. (2014). Un-sewered sanitation improvements for the urban-poor. African Water Facility publication. Online. Retrieved from https://www.africanwaterfacility.org/fileadmin/uploads/awf/Publications/Urban_Sanitation_Portfolio_Review.pdf.

Agyemang, I., McDonald, A., & Carver, S. (2007). Application of the DPSIR framework to environmental degradation assessment in northern Ghana. *Natural Resources Forum, 31*(3), 212–225.

Ajuang, C. O., Abuom, P. O., Bosire, E. K., Dida, G. O., & Anyona, D. N. (2016). Determinants of climate change awareness level in upper Nyakach Division, Kisumu County, Kenya. *Springer Plus, 5*(1), 1–20.

Asumadu-Sarkodie, S., & Owusu, P. (2016a). A review of Ghana's energy sector national energy statistics and policy framework. *Cogent Engineering, 3*, 1155274. https://doi.org/10.1080/23311916.2016.1155274

Asumadu-Sarkodie, S., & Owusu, P. A. (2016b). The potential and economic viability of solar photovoltaic in Ghana. *Energy Sources, Part A: Recovery, Utilization, and Environmental Effects*. https://doi.org/10.1080/15567036.2015.1122682

Asumadu-Sarkodie, S., & Owusu, P. A. (2016c). The potential and economic viability of wind farms in Ghana. *Energy Sources, Part A: Recovery, Utilization, and Environmental Effects*. https://doi.org/10.1080/15567036.2015.1122680

Asumadu-Sarkodie, S., Owusu, P. A., & Jayaweera, H. M. (2015a). Flood risk management in Ghana: A case study in Accra. *Advances in Applied Science Research, 6*, 196–201.

Asumadu-Sarkodie, S., Owusu, P. A., & Rufangura, P. (2015b). Impact analysis of flood in Accra, Ghana. *Advances in Applied Science Research, 6*, 53–78.

Barry, B., & Obuobie, E. (2012). Groundwater availability and use in sub-Saharan Africa: A review of 15 countries. In P. Pavelic, M. Giordano, B. Keraita, V. Ramesh, & T. Rao (Eds.). International Water Management Institute.

Bell, S. (2012). DPSIR a problem structuring method? an exploration from the "Imagine" approach. *European Journal of Operational Research, 222*, 350–360.

Bawakyillenuo, S. (2020). *Water pollution in Ghana*. Policy Brief. ISSER; P. O. Box.

Borja, A., Galparsoro, I., Solaun, O., Muxika, I., Tello, E. M., Uriarte, A., & Valencia, V. (2006). The European water framework directive and the DPSIR, a methodological approach to assess the risk of failing to achieve good ecological status. *Estuarine, Coastal and Shelf Science, 66*, 84–96.

Bunyasi, M. M. (2012). Vulnerability of hydro-electric energy resources in Kenya due to climate change effects: The case of the seven forks project. *Journal of Agricultural and Environmental Sciences, 1*, 36–49.

Caeiro, S., Mourao, I., Costa, M. H., Painho, M., Ramos, T. B., & Sousa, S., (2004). *Application of the DPSIR model to the Sado Estuary in a GIS context e social and economical pressures*. 7th AGILE Conference on Geographic Information Science 29 April–1 May 2004, Heraklion, Greece Parallel Session 4.3- Environmental/Social Modelling, 13.

Carr, E. R., Wingard, P. M., Yorty, S. C., Thompson, M. C., Jensen, N. K., & Roberson, J. (2007). Applying DPSIR to sustainable development. *International Journal of Sustainable Development and World Ecology, 14*(6), 543–555.

Carrier, M. A., Aubut, R., Lefebvre, R., Racicot, J., Asare, E. B., Gravel, E., & Rivera, A. (2011). *Hydrogeological atlas-Hydrogeological assessment project of the northern regions of Ghana*. Document submitted to the Canadian International Development Agency (CIDA) by SNC-Lavalin/INRS joint venture, December 2011, 58 p.

Central Intelligence Agency (CIA). (2021). Burkina Faso Details-The World Factbook. Retrieved April 22, 2021, from https://www.cia.gov/the-world-factbook/countries/burkina-faso/map.

Chacha, N., Njau, K. N., Lugomela, G. V., & Muzuka, A. N. (2018). Hydrogeochemical characteristics and spatial distribution of groundwater quality in Arusha well fields, Northern Tanzania. *Applied Water Science, 8*(4), 1–23.

CIAT, & World Bank. (2017). Climate-smart agriculture in Tanzania. *CSA Country Profiles for Africa Series, 53*(1).

Crawford, A., Price-Kelly, H., Terton, A., & Echeverría, D. (2016). *Review of current and planned adaptation action in Burkina Faso*. https://www.iisd.org/system/files/publications/idl-55876-burkina-faso.pdf

Crouzet, P., Beauzamy, B., & Carlier, C. (2009). Assessment of trends in water composition by stratification. Pilot application within the EEA DPSIR framework. *Houille Blanche-Revue Internationale De L Eau, 4*, 77–83.

EEA. (1999). *Environmental indicators: Typology and overview*. Technical report No 25. http://reports.eea.eu.int/TEC25/en/tab_content_RLR.

Elliot, M. (2002). The role of the DPSIR approach and conceptual models in marine environmental management: An example for offshore wind power. *Marine Pollution Bulletin, 44*.

Elisante, E., & Muzuka, A. N. (2017). Occurrence of nitrate in Tanzanian groundwater aquifers: A review. *Applied Water Science, 7*(1), 71–87.

FAO. (2020). *Aquastat Main Database*. Retrieved July 8, 2020, from http://www.fao.org/nr/water/aquastat/data/query/results.html.

FCFA. (2017). *Future climate projections for Tanzania*. Future Climate for Africa. www.futureclimateafrica.org/resource/future-climate-projections-for-tanzania/

Feldscher, K. (2018). *Students examine how climate change affects health*. Retrieved April 25, 2021, from https://www.hsph.harvard.edu/news/features/students-climate-change-health/

FEWS NET. (2018). Tanzania market fundamentals summary. *Office Markets & Public Policy*. https://doi.org/10.1002/9781118554302.ch2

Food and Agriculture Organization of the United Nations (FAO). (2021). *Action against desertification in the Burkina Faso*.

Ghiglieri, G., Pittalis, D., Cerri, G., & Oggiano, G. (2012). Hydrogeology and hydrogeochemistry of an alkaline volcanic area: The NE Mt. Meru slope (East African Rift–Northern Tanzania). *Hydrology and Earth System Sciences, 16*(2), 529–541.

GITEC, & WEMA. (2011). *Groundwater assessment of the Pangani Basin, Tanzania*. The Pangani Basin Water Board (PBWB) and the International Union for Conservation of Nature (IUCN) Report.

Global Facility for Disaster Reduction (GFDRR). (2011). climate risk and adaptation country profile Burkina Faso. https://climateknowledgeportal.worldbank.org/sites/default/files/2018-10/wb_gfdrr_climate_change_ country_profile_for_BFA.pdf.

Government of Kenya. (2007). *Kenya Vision 2030: A globally competitive and prosperous Kenya*. Government Printer (GoK).

Groen, J., Schuchmann, J. B., & Geirnaert, W. (1988). The occurrence of high nitrate concentration in groundwater in villages in northwestern Burkina Faso. *Journal of African Earth Sciences, 7*, 999–1009.

Herrero, M., Ringler, C., Van de Steeg, J., Thornton, P., Zhu, T., Bryant, E., Omollo, A., Koo, J., & Noternbaert, A. (2010). *Climate variability and change: Impacts on Kenyan Agriculture*. International Food Policy Research Institute, IFPRI.

Juma, D. W., Wang, H., & Li, F. (2014). Impacts of population growth and economic development on water quality of a lake: Case study of Lake Victoria Kenya water. *Environmental Science and Pollution Research, 21*(8), 5737–5746.

Kagalou, I., Leonardos, I., Anastasiadou, C., & Neofytou, C. (2012). The DPSIR approach for an integrated river management framework. A preliminary application on a Mediterranean Site (Kalamas River-NW Greece). *Water Resources Management, 26*, 1677–1692. https://doi.org/10.1007/s11269-012-9980-9

Kanoti, J. R., Olago, D., Opiyo, N., & Nyamai, C. (2019). An overview of groundwater and sanitation challenges in Kisumu City, Kenya. *International Journal of Research and Development*. https://doi.org/10.24940/ijird/2019/v8/i4/143592-347932-1-SM. ISSN 2278–0211 (Online).

Komakech, H. C., & de Bont, C. (2018). Differentiated access: Challenges of equitable and sustainable groundwater exploitation in Tanzania. *Water Alternatives, 11*(3), 623–637.

Komakech, H. C., & van der Zaag, P. (2011). Understanding the emergence and functioning of river committees in a catchment of the Pangani Basin, Tanzania. *Water Alternatives, 4*(2), 197–222.

Kristensen, P. (2004). *The DPSIR framework*. UNEP Headquarters, on a comprehensive/detailed assessment of the vulnerability of water resources to environmental change in Africa using river basin approach. Nairobi, Kenya.

Kuria, D. (2009). *Coping with climate change: Understanding local communities' knowledge and their coping strategies to climate change*. A project report to START.

Lin, T., Xue, X.-Z., & Lu, C.-Y. (2007). Analysis of coastal wetland changes using the DPSIR model: A case study in Xiamen, China. *Coastal Management, 35*(2), 289–303.

Martin, N., & van de Giesen, N. (2005). Spatial distribution of groundwater production and development potential in the Volta River basin of Ghana and Burkina Faso. *Water International, 30*(2), 239–249. https://doi.org/10.1080/02508060508691185

McCartney, M., Forkuor, G., Sood, A., Amisigo, B., Hattermann, F., & Muthuwatta, L. (2012). *The water resource implications of changing climate in the Volta River Basin*. International Water Management Institute.

Mdemu, M. V. (2021). Community's vulnerability to drought driven water scarcity and food insecurity in central and northern semi-arid areas of Tanzania. *Frontiers in Climate, 122*.

Ministry of Environment and Fishery Resources. (2015). Burkina Faso National Adaptation to Climate Change Plan (NAP).

MoALF. (2017). Climate risk profile for Kisumu County. Kenya County climate risk profile series. The Ministry of Agriculture, Livestock and Fisheries (MoALF). Prepared.

Molina, P. B., D'Alessandro, C., Dekeyser, K., & Marson, M. (2020). *Sustainable food systems through diversification and indigenous vegetables. An analysis of the Arusha area*. European Centre for Development Policy Management (ECDPM).

NEMA. (2011). *Effects of climate change and coping mechanisms*. State of the Environment Report Kenya 2006/7. Ministry of Environment and Natural Resources, Nairobi.

Njogu, A. K. (2000). *An integrated river basin planning approach–Nyando case study in Kenya*. 1st WARFSA/WaterNet Symposium: Sustainable Use of Water Resources; Maputo, 1–2 November 2000.

Notre Dame Global Adaptation Initiative (ND-GAIN). (2021). *Burkina Faso ND-GAIN Index*. Retrieved April 22, 2021, from https://gain-new.crc.nd.edu/country/burkina-faso.

Obuobie, E., Agyekum, W., Appiah-Adjei, E. K., Upton, K., Ó Dochartaigh, B. É., & Bellwood-Howard, I. (2018). *Africa groundwater atlas: Hydrogeology of Ghana*. British Geological Survey. Accessed [date you accessed the information]. http://earthwise.bgs.ac.uk/index.php/Hydrogeology_of_Ghana

Oikos, I. (2011). *The Mount Meru challenge: Integrating conservation and development in northern Tanzania*. Ancora Libri. www.istituto-oikos.org

Oiro, S. O. (2018). *The hydrogeology and groundwater quality assessment of Kisumu town area in Kenya*. Master Thesis, Ghent University, Gent, Belgium, 2012.

Olarinoye, T., Foppen, J. W., Veerbeek, W., Morienyane, T., & Komakech, H. (2020). Exploring the future impacts of urbanization and climate change on groundwater in Arusha, Tanzania. *Water International, 45*(5), 497–511.

Owen, M. O. (2020). Diagnostic assessment on availability and utilisation of water resources: A case study of Kisumu County and 20km adjacent area. *International Journal of Research and Scientific Innovation (IJRSI)*, 2321–2705.

Paavola, J. (2004). *Livelihoods, vulnerability and adaptation to climate change in the Morogoro region, Tanzania*. CSERGE Working Paper EDM, No. 04-12, University of East Anglia, The Centre for Social and Economic Research on the Global Environment (CSERGE), Norwich.

Pittalis, D. (2010). *Interdisciplinary studies for the knowledge of the groundwater fluoride contamination in the eastern African rift: Meru district-North Tanzania*. University of Sassari.

Potsdam Institute. (2020). *Climate risk profile: Burkina Faso*. https://www.pik-potsdam.de/en/institute/departments/climate-resilience/projects/project-pages/agrica/giz_climate-risk-profile_burkina-faso_en.

Relief Web. (2020). *Climate change and conflict pursue displaced Burkinabes-Burkina Faso*. https://reliefweb.int/report/burkina-faso/climate-change-and-conflict-pursue-displaced-burkinabes.

Sanon, V. P., Toé, P., Caballer Revenga, J., El Bilali, H., Hundscheid, L. J., Kulakowska, M., et al. (2020). Multiple-line identification of socio-ecological stressors affecting aquatic ecosystems in semi-arid countries: Implications for sustainable management of fisheries in sub-Saharan Africa. *Water, 12*(6), 1518.

Scheren, P. A. G. M. (2003). *Integrated water pollution assessment in data-and resource-poor situations: Lake Victoria and Gulf of Guinea case studies*. Technische Universiteit Eindhoven. https://doi.org/10.6100/IR564667.

SEI. (2009). *The Economics of Climate Change in Kenya*. Stockholm Environment Institute, SEI.

Shaghude, Y. W. (2006). Review of water resource exploitation and land use pressure in the Pangani river basin. *Western Indian Ocean Journal of Marine Science, 5*(2), 195–208.

Siwailam, M., Abdelsalam, H., & Saleh, M. (2019). Integrated DPSIR-ANP-SD framework for sustainability assessment of water resources system in Egypt. *IJAMSR, 3*(3).

Uhe, P., Sjoukje, P., Kew, S., Shah, K., Kimutai, J., Mwangi, E., van Oldenborgh, G. J., Singh, R., Arrighi, J., Jemba, E., Cullen, H., & Otto, F. (2016). Attributing drivers of the 2016 Kenyan drought. *International Journal of Climatology, 38*, 554–568.

USAID. (2012). Climate risks in food for peace and geographies: Burkina Faso. Water and sanitation in Burkina Faso (2011). Turning finance into services from 2015 and beyond. Water and Sanitation Program–Africa Region (2011). The World Bank.

Wagah, G., Onyango, M. G., & Kibwage, J. K. (2010). Accessibility of water services in Kisumu municipality, Kenya. *Journal of Geography and Regional Planning, 3*(4), 114–125. http://www.academicjournals.org/JGRP. ISSN 2070-1845 © 2010 Academic Journals

Walsh, G. (2019). *Responding to drought, hunger and climate change in Kenya*. Retrieved January 26, 2021, from https://reliefweb.int/report/kenya/responding-drought-hunger-and-climate-change-kenya#:~:text=Kenya is experiencing severe drought, with one meal a day.&text=The dry and arid Turkana, drought this year in Kenya.

Wasonga, J., Okowa, M., & Kioli, F. (2016). Sociocultural determinants to adoption of safe water, sanitation, and hygiene practices in Nyakach, Kisumu County, Kenya: A descriptive qualitative study. *Journal of Anthropology*.

Yameogo, S., & Savadogo, A. N. (2002). Les Ouverages de Captage de la ville de Ouagadougou et LeurVulnerabilite a la Pollution. In A. H. Maiga, L. S. Pereira, & A. Musy (Eds.), *Sustainable water resources management: Health and productivity in hot climates*.

Yeleliere, E., Cobbina, S. J., & Duwiejuah, A. B. (2018). Review of Ghana's water resources: The quality and management with particular focus on freshwater resources. *Applied Water Science, 8*(3), 1–12.

A Review of Guidelines on Agri-food Value Chain Modeling

H. Ece Salali, Yarkın Akyüz, Pelin Atakan, Cihat Günden, and Murat Yercan

1 Introduction

Value chains have been appropriate for development in agricultural and food systems because much of the work on VC development started in agriculture. It is important to understand the impact of the VC on the structure and condition of the incomes and working conditions of those agents involved, and on food security and environmental sustainability. Governments and non-governmental organizations (NGOs) use VC development for promoting economic growth and combating rural poverty. Hence, there has been an increasing development on guidelines. It is thought that by focusing on the VC and the links between the actors, the problems among actors in the chain can be identified. Thus, the objective of this study is to examine some well-known and widely used guidelines of agri-food VCs in order to identify and review the concept of agri-food VC tools.

The value chain is a full range of activities required to bring together a product or service through the different phases of processes in order to deliver to the final consumers, with disposal after use. Furthermore, a VC exists when all of the agents in the chain operate in a way that maximizes the generation of value along the chain (Kaplinsky, 1999). These definitions can be communicated in narrow or broad approaches, which are given as follows:

Firstly, a VC focuses on a single firm and includes the conception and design stage; the acquisition of inputs; production, marketing, and distribution activities; and the performance of after-sales services in the narrow approach.

H. E. Salali (✉) · Y. Akyüz · C. Günden · M. Yercan
Faculty of Agriculture, Department of Agricultural Economics, Ege University, İzmir, Turkey

P. Atakan
Graduate School of Natural and Applied Sciences, Department of Agricultural Economics, Ege University, İzmir, Turkey

K. Mattas et al. (eds.), *Value Chain Dynamics in a Biodiverse Environment*, Cooperative Management, https://doi.org/10.1007/978-3-031-49845-9_6

Secondly, the broad approach of defining a VC is the complex range of activities implemented by various agents (primary producers, processors, traders, service providers) to bring a raw material through a chain to the sale of the final product. The "broad" VC starts from the production system of the raw materials and will move along the linkages with other enterprises engaged in trading, assembling, processing, etc. The broad approach does not only include the activities implemented by a single enterprise, but also all its backward and forward linkages, until the level in which the raw material is produced will be linked to the final consumers (BIOVALUE – D2.1, 2022). The broad concept of VC also contains the issues of organization and coordination, the strategies, and the power relationships of the different actors in the chain. It is important to understand that conducting a VC analysis requires a thorough investigation of what is going on between the agents in a chain, what keeps these agents together, what information is shared, and how the relationships between the agents are evolving. In addition, the idea of broad concept of VC is associated with the concept of governance, which is of key importance for those researchers interested in the social or environmental facets of VC analysis. The establishment of VCs may put pressure on natural resources (such as water or land) which may produce degradation of the soil, loss of biodiversity, or pollution. Conditioned chain relations and performance are expected for increasing the competitiveness of the agricultural sector. Individual actors such as smallholders and cooperatives are important for achieving development goals.

Important factors that have spurred interest in VC development include a growing demand for added-value foodstuffs in developing countries, more food safety standards by governments and private firms, the growth of niche markets (e.g., organic and fair trade), and concern over the scarcity of agricultural raw materials (Donovan et al., 2015). The rapid growth in demand for agrifood products in which smallholders are considered to have a comparative advantage—for example, specialty crops that require high labor inputs—has been considered an opportunity to combine economic growth and poverty reduction goals (Bacon, 2005; Weinberger & Lumpkin, 2007).

As an analytical framework, the VC concept serves as a way to explore the actors, structures, and dynamics of VCs, focusing on the positioning of chain actors, the linkages between chain actors, the distribution of value added along the chain, and opportunities for upgrading. *Therefore, the major problem is to select the appropriate VC guide that meets the four-dimensional approach with strong potential for accelerating the objectives of the BIOVALUE project.*

Value chains can be divided into groups according to the approach they used: Activity-based VC analysis is based on all related main and supportive activities in the VC from the first step to the end market, such as input provision, cultivation, collection, production, marketing, and distribution and consumption. Agent-based VC analysis is based on the individual or institutional actors dealing with the activities done in the VC, such as farmers, collectors, processors, wholesalers, retailers, and consumers.

In practice, the scope of VCs takes different forms. First, the most common and longest in use is the "pro-poor VC" approach which reflects an understanding that market integration for poor farmers or input and service providers can be achieved by reducing the barriers to enter markets. Second, "nutrition-sensitive VC" approaches have gained increased popularity attempting to improve nutritional outcomes along with promoting increases in incomes for smallholders by alleviating constraints in the supply or demand of foods. Third and most recently, the "greening of VCs" approach has received increased interest, particularly in response to climate change and sustainability challenges. Hence, the guidelines differ in their developmental approach, their developmental goals, and their targeted users. Guidelines also vary in terms of their information requirements, objectives and overall complexity, conceptualization of VC concepts, and incorporation of local actors into research and strategy formulation, among other factors (Donovan et al., 2015).

Various guidelines exist for carrying out VC analysis. Nang'ole et al. (2011) have investigated VC manuals, VC guidelines, and VC handbooks. Their study starts by reviewing the commonalities and differences in the definition of VC and other relevant terms. Four stages of VC analysis are described: appraisal, design, implementation and monitoring, and evaluation.

Donovan et al. (2015) have reviewed 11 guidelines for VC development. They have reviewed the current guidelines on the basis of VC objectives and motivations, VC definitions, methodological design of VC, data collection and analysis, and assessing and monitoring outcomes and impacts.

Clay and Feeney (2019) have done a literature review on analyzing agribusiness VCs. Their study is divided into two parts. The first part of the analysis delves into the VC concept, with the aim of discovering how the concept has changed from a historical perspective, and what would be an acceptable narrow definition. The second and main part of the analysis seeks to study the methodological techniques for approaching and accurately analyzing a VC in the agribusiness sector.

Stoian et al. (2012) present a strong case for why those who aim to advance inclusive VC development need to better consider the needs and circumstances of the rural poor. They emphasize the bottlenecks, trade-offs, and dilemmas that can arise when attempting to link poor farming households with higher-value markets. This study seeks an answer to the question about which guidelines provide an insight for sustainable food systems that require different approaches to ensure economic and social prosperity while preserving the environment and biodiversity for agri-food VC.

Guidelines for VC are an analytical tool to design interventions on behalf of smallholders and small rural businesses that are affected by the expansion of international agribusiness (Haggblade, 2007). Thus, public and private sector leaders, stakeholders, NGOs, and researchers can advance the selected VCs for more sustainable and resilient agro-food channel systems for smallholder producers.

2 Materials and Methods

In order to identify and review the agri-food VC tools, the study proceeded first with the theoretical information including some important well-known, widely used guidelines of VC analysis. The guidelines were used to produce data on the tools/methodologies/practical approaches of VC analysis and the indicators/outcomes could be used for each related tool where each one has some specific objectives. This is called a guideline-based review.

Fourteen well-known and widely used guidelines have been reviewed based on their tested concepts and methods and the tools used and endorsed with their case study projects. The review process is based on the guidelines and on the data which are tested exactly and approved by the case study results. All investigated guidelines have been reviewed on the basis of the tools used which means the area of VC analysis and the produced outcomes in each related tool.

3 Results

The results of the review based on the VC guidelines are presented under two headings: selection and comparison. After that, the determined indicators and outcomes are also presented.

Table 1 Reviewed VC guidelines

The Reviewed Guidelines	
ILO	Value Chain Development for Decent Work (2021)
VCA4D	Value Chain Analysis for Development (2018)
ACIAR	Australian Center for International Agricultural Research (2016)
GTZ/GIS	Guidelines for Value Chain Selection (2015)
FAO	Developing Sustainable Food Value Chains (2014)
FAO	VC Analysis for Policy Making (2013)
UNIDO	United Nations Industrial Development Organization (2011)
IIED	International Institute for Environment and Development (2008)
M4P	Making VCs Work Better for the Poor (2008)
USAID	United States Agency International Development (2008)
GFU	Promoting Value Chains of Neglected and Underutilized Species (2008)
CIAT	Centro Internacional de Agricultura Tropical (2007)
FAO	Rapid Appraisals (2007)
CIP	International Potato Center (2006)

3.1 Selecting the VC Guide

The study is started first with the review of the guidelines for the VC analysis. The results showed that the 14 different guidelines listed below are well known and frequently used for VC analysis (Table 1).

ILO—Value Chain Development for Decent Work VC Development for Decent Work is a guide that takes a systems approach to VC development with the goal of creating more and better jobs. The guide is based on the International Labour Organization's (ILO) vast experience in using the systems approach for VC development and focuses on decent work outcomes across the four pillars of the ILO Decent Work Agenda.

VCA4D—Value Chain Analysis for Development The purpose of VC Analysis for Development (VCA4D) is to provide decision-makers with evidence-based information to feed sustainable development strategies. It is directed at policy makers and stakeholders, and in this regard aligns with the EU aims as an aid provider, and fits within its policy dialogue approach. Analyzing VCs sheds light on impacts, uncovers main pathways, and identifies at which stages of the chain and for which actors investment and support can generate benefits, eliminate drawbacks and constraints, and foster sustainability and inclusiveness.

ACIAR—Australian Center for International Agricultural Research The ACIAR manual tries to promote "value chain thinking," which means taking a whole-of-chain perspective, emphasizing the importance of understanding markets and consumers, and collaboration among chain members. It highlights how effective partners can align their skills, resources, and behavior to deliver products and services to receptive consumers and to reduce waste, with the resultant financial returns being distributed equitably so as to sustain the partnerships. It helps chain members to recognize their interdependence, and the consequent benefits of building collaborative relationships for solving the shared problems of creating and delivering consumer value (Collins et al., 2016).

GIZ/GTZ—Guidelines for VC Selection Economic, Environmental, Social and Institutional Guidelines for VC Selection are based on development practitioners working in VC development, specifically with GIZ and the ILO. The document includes criteria and tools for VC selection with the goal of generating the greatest impact in accordance with specific development objectives and project mandates through interventions to be designed within the project. It takes a holistic approach to the VC selection process by integrating four dimensions as follows: economic, environmental, social, and institutional (Schneemann & Vredeveld, 2016).

FAO—Developing Sustainable Food Value Chains The SFVC concept recognizes that VCs are dynamic, market-driven systems in which vertical coordination (governance) is the central dimension and for which value added and sustainability are explicit, multidimensional performance measures, assessed at the aggregate level. SFVC is a market-oriented and systems-based approach for measuring, analyzing, and improving the performance of food VCs.

FAO—VC Analysis for Policy Making (Quantitative Approach for the Policy Impact Assessment) This guide is developed by the FAO, which is especially focused on policy making through the quantitative approach. These guidelines provide users with the key notions required to carry out analyses of policy impacts by means of a VC approach and show how to do it by making use of relevant approaches and tools. In particular, users will find this material useful to identify the main features of a given VC and to build consistent VC accounting frameworks. Besides, building alternative scenarios reflecting changes that given policy measures are likely to introduce in VCs are also useful for the users. They are also useful to measure in monetary terms shifts in physical production, value added, and income accruing to the various agents involved, and they provide relevant information to decision-makers and other stakeholders involved in policy-making processes, which makes this guide useful. For instance, the user will be driven to identify the basic units operating in a given VC and the activities they undertake, quantifying the revenue, value added, and profits of every agent, building different scenarios for selected policy options, calculating value added and other margins, and computing protection and competitiveness indicators (Bellu, 2013

UNIDO—United Nations Industrial Development Organization The guide builds on a review of common practices in VC development projects in Asia and the Pacific region as well as on experience from six case studies of VC development projects in Sri Lanka, Vietnam, and Indonesia.

IIED—International Institute for Environment and Development (A Guide to Multi-Stakeholder Process for Linking Small-Scale Producers to the Modern Market) This guide has been developed through a collaborative process involving many of the partners from within the regoverning markets and consortia. Broadly, this guide is for anyone interested in finding practical ways to enhance opportunities for small-scale producers in modern markets. Users may be market actors interested in creating direct links with small-scale suppliers, government policy makers tasked with rural development, producer organizations working for their members, NGOs working for the rural poor, or researchers working to understand and support processes aimed at greater inclusion of small-scale producers.

M4P—Making VCs Work Better for the Poor (Poverty Reduction Impact of VC for the Poor) This guide provides VC practitioners with an easy-to-use set of tools for VC analysis, with a focus on poverty reduction. The aim of this guide is to strengthen the links between VC analysis and development interventions that improve the opportunities available to the poor (Anonymous, 2008). The guide is designed as a concise manual to be used in the field and by those involved in project development and/or assessment of investment opportunities. The focus is on providing easy to follow tools and clear explanations about their use. This includes examples of how these can and have been used in real VC analyses in the past. Although the VC analysis theory that underpins the tools presented in the guide is an important element, the practical aspects of analysis dominate the guideline content (Anonymous, 2008).

USAID—United State Agency International Development (End Market Research Toolkit Upgrading VC Competitiveness with Informed Choice) The success of private firms is only the beginning of successful VC development. Successful VC development involves the creation of a vibrant VC where all stakeholders are focused on the needs of the market and create collaborative business models that promote equitable growth. Achieving this vision of broad-based economic growth is the ultimate goal of VC development. End market research should be the first step in designing a competitiveness strategy that creates a roadmap for identifying and serving the best customers in the world for the products and services that developing country VCs are able to sell.

GFU—Promoting VCs of Neglected and Underutilized Species This guide presents stages and good practices for VC development of neglected and underutilized species (NUS). The guide begins first by giving a brief introduction to basic concepts for VC development of neglected and underutilized species, and the impact of neglected and underutilized species in VC development striving for social, economic, and environmental impacts, drivers fostering and hampering the utilization of biodiversity. This is followed by an introduction of the strategic cycle for participatory VC development. The fourth chapter takes stock of methodologies and tools for building structure and capacities for sustainable neglected and underutilized VC development. Building on these guiding principles and possible approaches to NUS-VCD, the last two chapters look at the questions of how far and with what preconditions NUS-VCD can contribute to the main objectives of biodiversity conservation and pro-poor growth by discussing social, economic, and environmental impacts and summarizing lessons learnt from case studies and other fields of experiences.

CIAT—Centro International Agricultural Development (Participatory Market Chain Analysis for Smallholder Producers) Among the first guidelines available for helping development practitioners work with smallholders and small businesses in formulating a VCD strategy, it presents a relatively simple and well-integrated conceptual framework. Implementation is designed to be highly participatory.

FAO—Rapid Appraisals This guideline for rapid appraisals of agri-food chain performance in developing countries presents a methodological strategy for the analysis of agri-food VCs. Simply stated, chains can be seen as sets of interrelated activities that are typically organized as sequences of stages. In the agricultural, food, and fiber sector, chains encompass activities that take place at the farm level, including input supply, and continue during first handling, processing, and distribution.

CIP—International Potato Center (Participatory Market Chain—Qualitative Approach) The main objective of this guide is to present the participatory market chain approach (PMCA), which aims to stimulate market chain innovations by involving different stakeholders within a well-structured and demand-oriented process. By presenting both theory and practice, this guide should enable the leaders to apply the PMCA method in the specific context in which they are working.

Each guideline has some specific tools and indicators because of their specific objectives such as regional development, rural development, product development or product replacement, and decent work analysis in a sector. These are ILO: VC development for decent work; VCA4D: VC analysis for development; FAO: developing sustainable food VC; FAO: VC analysis for policy making; M4P: making VC work better for poor; GFU: promoting VCs of neglected and underutilized species; UNIDO: pro-poor VC development; and some others. It has been seen that a VC analysis can concentrate on four different dimensions, which are institutional/functional analysis, economic/financial analysis, social analysis, and environmental analysis. Some guidelines have a full range of interest with the four dimensions, but some of them have fewer areas of interest. Also, the guidelines make their analysis by using methodology on activity-based, agent-based, or both approaches. The activity-based approach includes all activities in the core process from the first step to the last step of the VC. The agent-based approach includes all agents involved along the VC from up- and downstream (Table 2).

On the other hand, the review process includes the area of interest of modern food VCs and their linkages with the biodiversity, potential reasons, causes, and conditions, for abandonment of certain species cultivations and consumption patterns of healthy and environmentally friendly foods.

Value chain analysis consists of two different approaches: activity-based and agent-based. Activity-based VC analysis is based on the all related main and supportive activities in the VC from the first step to the end market, such as input provision, cultivation, collection, production, marketing, and distribution and consumption. Agent-based VC analysis is based on the individual or institutional actors dealing with the activities done in the VC, such as farmers, collectors, processors, wholesalers, retailers, and consumers. In Table 2, the guidelines were given with their concentrated subjects, the survey instruments used, and the types of approaches used. All these guidelines are developed by different institutes and tested for different countries and crops. The guidelines included here are broad enough to provide a strong indication of the overall state of the art. The review process is based on the guidelines themselves and based on the data on which they are tested exactly and approved by the case study results.

Past development operations frequently focused on increasing agricultural produce while often ignoring the market and the other economic drivers involved. Production activities are part of a wider network of interdependent businesses, and

Table 2 Reviewed guidelines according to the concentrated subject, survey instrument, and approach used

Guidelines	Concentrated subject	Survey instrument	Approach used	
			Activity based	Agent based
ILO—Value Chain Development for Decent Work[1]	Working conditions, social welfare, and legal protection of employees throughout a VC	Conducting individual survey with agents, and collecting of secondary data on legislation and practices	x	x
VCA4D—Value Chain Analysis for Development[2]	Economic, Social, and Environmental Analysis	Secondary data analysis, focus group meeting, interview with stakeholders, software needs	x	x
ACIAR—Australian Center for International Agricultural Research[3]	A guide to VC analysis and development for overseas development assistance projects	Workshops and focus group meeting, desktop studies, consumer research, gender analysis		x
GTZ/GIS—Guidelines for Value Chain Selection[4]	Economic, environmental, social, and institutional	Secondary data analysis, key informants' interviews	x	
FAO—Developing Sustainable Food Value Chains[5]	Economic, social, and environmental impact of VC	Secondary data analysis, questionnaire-based survey with stakeholders, software needs		x
FAO—VC Analysis for Policy Making[6]	Quantitative approach for the policy impact assessment	Secondary data analysis, statistical databases, focus group discussion, semi-structured interviews and questionnaire		x
UNIDO—United Nations Industrial Development Organization[7]	Pro-poor VC development-functional and social VC	Secondary data analysis, stakeholder interview		x
IIED—International Institute for Environment and Development[8]	A guide to multi-stakeholder processes for linking small-scale producers to modern markets	Participatory process, stakeholder workshops	x	
M4P—Making VCs Work Better for the Poor[9]	Poverty reduction impact of VC for the poor	Key informant interviews, secondary data analysis	x	x
USAID—United States Agency International Development[10]	End market research toolkit upgrading VC competitiveness with informed choice	Secondary data analysis, focus group meeting, consumer survey		x

(continued)

Table 2 (continued)

Guidelines	Concentrated subject	Survey instrument	Approach used	
			Activity based	Agent based
GFU—Promoting Value Chains of Neglected and Underutilized Species[11]	Try to promote neglected and underutilized species' VCs	Rapid appraisal, key informants' interviews, in-depth surveys, desk survey		x
CIAT—Centro Internacional de Agricultura Tropical[12]	Participatory market chain analysis for smallholder producers	Secondary data analysis, key informants' interviews, focus group discussion	x	x
FAO—Rapid Appraisals[13]	Guidelines for rapid appraisals of agri-food chain performance in developing countries	Key informants' interviews, structured direct observations		x
CIP—International Potato Center[14]	Participatory market chain—Qualitative approach	Rapid market appraisal, focus group, and quantitative market study		x

Sources: (1-ILO, 2021; 2-EC, 2018; 3-Collins, et.al., 2016; 4-Schneemann & Vredeveld, 2016; 5-FAO, 2014; 6-Bellu, 2013; 7-UNIDO, 2011; 8-Vermeulen, et.al., 2008; 9-Anonymous, 2008; 10-Henning, et.al., 2008; 11-Will, 2008; 12-Lundy, et.al., 2007; 13-Silva & Souza, 2007; 14-Bernet, et.al., 2006)

it is therefore essential to examine them within the VC as a whole. Moreover, interventions in agriculture seldom paid enough attention to the related environmental and social impacts. Yet, decision-makers must think over the fact that VC activities take place in and influence a social and environmental context.

Other modeling tools have been investigated on production processes, product delivery channels, price transmission, quality controls, and investment planning. The production process is investigated with the data on up-mid-downstreams of VCs and these are comprised of cases on livestock breeding, seed growers, horticultural production, dairy products, the grain industry, the fruit and vegetable processing industry, the meat processing industry, and post-processing stages like retailers and consumers. Product delivery channels have investigated the multiple stops from the production phase to the marketplace, where the product is sold to the end consumer. The characteristics of products such as perishability, size, and unit value, the necessary storage conditions, the distance between the production area and the end customers, competition, buying capacity of intermediaries, number of intermediaries, end market size, competition, and environmental concerns are among the factors that shape the choice of distribution channels.

Fourteen well-known and widely used guidelines have been reviewed based on their concepts and methods tested and tools used and endorsed with their case study projects. All investigated guidelines have been reviewed on the basis of the tools used which means the area of VC analysis and the produced outcomes in each related tool. The 14 guidelines are investigated through the area of interest which concentrates on topics according to the analytical framework of the VC analysis. After reviewing the guidelines, it is required to compare each guideline by their dimensions.

3.2 Comparing VC Guidelines

All the guidelines are investigated through the area of interest which concentrates on topics according to the analytical framework of the VC analysis. The area of interest of the guidelines differs from one to the other through the expectations and objectives of the project that they are targeting. The guidelines are investigated on topics according to the analytical framework of the VC analysis such as institutional/functional, economic/financial, social, and environmental (Table 3).

Institutional/functional analysis provides a detailed profile of the industry structure through the identification, description, and quantification in physical terms of the sequence of operations concerning commodity production, processing, marketing, and final consumption. Institutional/functional tools are mapping, governance analysis, demand and supply conditions, SWOT analysis, and end market analysis.

The economic approach of a VC assesses in quantitative terms the creation of value added and its distribution to the various agents involved. Other tools are the policy analysis matrix with financial ratio analysis.

The social approach consists of the tools of employment creation, gender analysis, gender equality, and decent work deficit evaluation.

The evaluation of the environmental approach of the VC is made by hot spot analysis, environmental assessment, and life cycle assessment, which have direct effects, decreasing or increasing the biodiversity.

3.3 Outcomes Produced by the Guidelines

The evaluation process of the different guidelines on agri-food VC analysis concentrates on four different dimensions in a broader approach. But most of the guidelines run with some part of them in a particular approach, mainly with the institutional/ functional and economic/financial tools.

3.3.1 Institutional/Functional Analysis

Institutional/functional analysis provides a detailed profile of the industry structure through the identification, description, and quantification in physical terms of the sequence of operations concerning commodity production, processing, marketing, and final consumption. More specifically, it examines:

Table 3 Guidelines by their dimensions

Guidelines	Institutional/ functional analysis	Economic/ financial analysis	Social analysis	Environmental analysis
ILO—Value Chain Development for Decent Work (2021)		x	x	
VCA4D—Value Chain Analysis for Development (2018)	x	x	x	x
ACIAR—Australian Center for International Agricultural Research (2016)	x	x	x	x
GTZ/GIS—Guidelines for Value Chain Selection (2015)	x	x	x	x
FAO—Developing Sustainable Food Value Chains (2014)	x	x	x	
FAO—VC Analysis for Policy Making (2013)	x	x		
UNIDO—United Nations Industrial Development Organization (2011)	x		x	
IIED—International Institute for Environment and Development (2008)	x			
M4P—Making VCs Work Better for the Poor (2008)	x	x	x	
USAID—United States Agency International Development (2008)	x	x		
GFU—Promoting Value Chains of Neglected and Underutilized Species (2008)	x	x	x	
CIAT—Centro Internacional de Agricultural Tropical (2007)	x			
FAO—Rapid Appraisals (2007)	x			
CIP—International Potato Center (2006)	x	x		

Sources: (ILO, 2021; EC, 2018; Collins, et.al., 2016; Schneemann & Vredeveld, 2016; FAO, 2014; Bellu, 2013; UNIDO, 2011; Vermeulen, et.al., 2008; Anonymous, 2008; Henning, et.al., 2008; Will, 2008; Lundy, et.al., 2007; Silva & Souza, 2007; Bernet, et.al., 2006)

(a) Technical operations required from primary production to final consumption
(b) Inputs used and intermediate outputs produced at each stage of the chain
(c) Economic agents involved at the different stages and related functions
(d) Physical flows of the commodity among the different agents

(e) Bottlenecks (e.g. inputs availability, and logistical issues)

This is called mapping in general. Others come behind mapping: governance analysis, demand and supply conditions, SWOT analysis, and end market analysis. Each one of them has some specific data produced (Table 4).

3.3.2 Economic/Financial Analysis

The "economic analysis" of a VC assesses in quantitative terms the creation of "value added" and its distribution to the various agents involved. The value added is a measure of wealth created in an economic system by a production process, the net of the resources consumed by the process itself. More specifically, the economic analysis allows the analyst to determine:

(a) The value added created by the overall VC
(b) The value added and margins for each economic agent at each stage of the chain
(c) The allocation of value added among production factors (capital labor, other assets) and the public budget, through the respective distributive variables (profits, wages, rents, and taxes)
(d) Providing some data for investment planning and financial analysis that can be ensured by the value-added data that resulted.

Another tool is to prepare a policy analyses matrix (PAM). The fourth tool (end market analysis) which is a very important part of the VC analysis contains some explanatory data on consumer behavior (Table 5).

3.3.3 Social Analysis

Social analysis of the VC consists of the contribution of the VC to the socio-economic situation, including income, expenditure, and other social well-being implications for various social groups of interest to the VC, such as the geographic location of the VC and implications for territorial set-up and development (rural–urban relationships, synergies with other activities, role in local production systems, etc.). This analysis consists of the elements of employment creation, gender analysis and gender equality, and decent work deficit evaluation (Table 6).

3.3.4 Environmental Analysis

The result of the environmental analysis of the VC is done to prove the increase or no increase in biodiversity. The evaluation of the environmental side of the VC is made by hot spot analysis, environmental assessment, and life cycle assessment (Table 7).

Table 4 Outcomes of institutional/functional analysis

Tools	Indicators/outcomes
Mapping (VC element analysis)	1. Functional analysis table 2. Mapping of core process 3. Mapping of agents 4. Mapping flows of products 5. Marketing channels 6. Volume of inputs/outputs 7. Mapping knowledge and flow of information 8. Mapping the volume of products, number of agents and jobs 9. Mapping the value at different levels of the VC 10. Mapping the relations, linkages, and trust 11. Mapping constraints and potential solutions 12. Knowledge and technology matrix 13. Matrix of market channel analysis
Governance analysis	1. Matrix of rules, standards, and regulations 2. Matrix of regulations and agents 3. Quality standards 4. Rewards and functions 5. Access to market, technologies, finance, skills, and knowledge 6. Vertical-horizontal integration 7. List of constraints (and type of constraints) 8. List (and type) of relevant economic support programs running and planned for 9. Certification/labeling
Demand and supply conditions	1. Area sown, yields, number of farmers (time series) 2. Quantity of supply and demand (time series) 3. Export and Import (time series) 4. Prices (national market, international) (time series) 5. Supply Utilization Account (time series) 6. Share (%) of product/sector in Gross Domestic Production (GDP) and export value
SWOT analysis	1. Strengths of VC 2. Weaknesses of VC 3. Opportunities of VC 4. Threats of VC
End market analysis (market research/Phase 1)	1. National market 2. International market (import-export)

Sources: (EC, 2018; Collins, et.al.,2016; Schneemann & Vredeveld, 2016; FAO, 2014; Bellu, 2013; UNIDO, 2011; Vermeulen, et.al., 2008; Anonymous, 2008; Henning, et.al., 2008; Will, 2008; Lundy, et.al., 2007; Silva & Souza, 2007; Bernet, et.al., 2006)

4 Conclusions

This study has examined a guideline review process for agri-food VCs. The process required a step-by-step examination for assessing, comparing, and selecting the VCs. The study determines different approaches that promote selecting the VCs by using secondary data. It was identified that VCs based on the institutional/functional, economic/financial, social, and environmental approach are promoting the

Table 5 Outcomes of economic/financial analysis

Tools	Indicators/outcomes
Value-added analysis	1. Total output value 2. Cost of intermediate inputs 3. Cost of fixed capital 4. Gross VA 5. Net VA 6. Income distribution by sources and agents 7. Competitiveness
Financial analysis	1. Cost benefit (CB) 2. Net present value (NPV) 3. Internal rate of return (IRR) 4. Payback period 5. Total output value 6. Cost of intermediate inputs 7. Cost of fixed capital 8. Cash flows 9. Break-even point
PAM (policy analysis matrix)	1. Private cost ratio (PCR) 2. Private VA ratio 3. Domestic resource cost (DRC) 4. Social VA ratio 5. Nominal protection coefficient (NPC) 6. Effective protection coefficient (EPC) 7. Domestic factor ratio (DFR)
End market analysis (psychographic analysis of farmers/consumers/ Phase 2)	1. Value 2. Attitude 3. Behavior 4. 4) Preferences

Sources: (ILO, 2021; EC, 2018; Collins, et.al., 2016; Schneemann & Vredeveld, 2016; FAO, 2014; Bellu, 2013; Anonymous, 2008; Henning, et.al., 2008; Will, 2008; Bernet, et.al., 2006)

Table 6 Outcomes of social analysis

Tools	Indicators/outcomes
Employment created	1. Labor needs by agents-activities-process 2. Wage differentiation 3. Labor intensity: number of persons employed in various VC stages
Gender analysis	1. Women and youth participation 2. Number of persons (M/F) employed in the value chain (sector) and trends
Decent work deficit analysis	1. List and level of health and safety risks. Incidence of occupational accidents in the workplace; working time lost due to sickness; worker perceptions of physical and mental well-being 2. Job security and safety: type of employment (contractual status, legal benefits of the contracts, duration of contracts); presence of precarious conditions; presence of additional disadvantages due to gender, ethnicity, or race.

Sources: (ILO, 2021; EC, 2018; Collins, et.al., 2016; Schneemann & Vredeveld, 2016; FAO, 2014; UNIDO, 2011; Anonymous, 2008; Will, 2008)

selection of VCs, which have strong potential for accelerating the expectations and objectives of the study. These are mostly depending on the guidelines' objectives. Hence, this review consists of an approach with the four dimensions of the VC analysis: institutional/functional, economic/financial, social, and environmental evaluation. Institutional/functional analysis provides a detailed profile of the industry structure through the identification, description, and quantification in physical terms of the sequence of operations concerning commodity production, processing, marketing, and final consumption.

The institutional/functional analysis aims to build an overall description of the VC system. It identifies and represents the main actors and stakeholders involved and expands on some of the main strategic development challenges faced. This analysis can produce important data on the production process, up-down streams, product delivery channels, and quality control schemes. The main tool of this analysis is mapping which is the first step of the analysis. The other sides of the institutional/functional analysis are governance analysis, demand and supply analysis, SWOT analysis, and end market analysis. Institutional/functional analysis of the

Table 7 Outcomes of environmental analysis

Tools	Indicators/outcomes
Hot spot analysis	1. Material consumption 2. Energy consumption 3. GHG Emissions 4. Water consumption 5. Land (erosion, pollution) 6. Air pollution 7. Water pollution 8. Waste 9. Biodiversity 10. Impact of environmental degradation on the VC
Environmental assessment	1. Resource depletion 2. Ecosystem quality 3. Human health 4. Biodiversity loss
Life cycle assessment	1. Climate change—total, fossil, biogenic, and land use 2. Ozone depletion 3. Acidification 4. Eutrophication—freshwater 5. Eutrophication—marine 6. Eutrophication—terrestrial 7. Photochemical ozone formation 8. Depletion of abiotic resources—minerals and metals 9. Depletion of abiotic resources—fossil fuels 10. Human toxicity cancer, non-cancer 11. Eco-toxicity (freshwater) 12. Water use 13. Land use 14. Ionizing radiation, human health 15. Particulate matter emissions

Sources: (EC, 2018; Collins, et.al., 2016; Schneemann & Vredeveld, 2016)

VC produces data on the production process and the up- and downstreams of the VC, the data on quality control schemes, price transmission, product delivery channels, cultivation, and the consumption patterns of the VCs.

The economic/financial analysis aims at measuring and interpreting the profitability and sustainability of the VC operations for all the actors directly involved. Its purpose is to inform on the economic effects of the VC within the national economy in terms of growth generation and distribution of incomes. It also assesses its competitiveness and viability within the global economy. Thus, generally, this analysis uses monetary-equivalent data. This analysis can produce important data on investment planning to be foreseen and the data on farmers' and consumers' psychographic analysis with their values, attitudes, behaviors, and preferences.

Social analysis is concentrated on two dimensions. First is the social inclusiveness of the VC. This part of the analysis elaborates how the VC organization and governance involve the various stakeholders and how the incomes and employment generated are distributed among social groups. The VC has some specific impacts on vulnerable groups such as subsistence-oriented farmers, smallholders, women, youth, and marginalized people (landless rural workers, minority communities). The second issue is social sustainability. The analysis of social sustainability focuses on assessing established and potential consequences of the VC operations in an array of six domains of importance for decision-makers because they convey key concerns of development: working conditions, land and water rights, gender equality, food and nutrition security, social capital, and living conditions.

The environmental analysis of the VC mainly relates to "resource depletion," "ecosystem quality," "human health," "climate change," and "biodiversity." By combining data and findings on the various areas related to the topics above, qualitative and quantitative appraisal has to be done for the environmental sustainability of the VC. The approach to evaluate the environmental sustainability of the VC is twofold, based on the quantitative life cycle assessment (LCA) accompanied by an exploratory assessment of biodiversity risks.

Based on differentiating between two approaches used: activity- and agent-based selection, it is important for the subsequent step after focusing on agricultural and food VCs. The review results indicate that the researchers mainly studied institutional/functional analysis which is the first step of VC analysis. Another most studied area of interest is the economic/financial analysis of VC. The social and environmental sides of the VC have been studied less. It is underlined that researchers generally used mapping of the VC as it is an important area of interest and the first step for the VC analysis. The other tools are demand and supply analysis, governance analysis, and end market analysis of VCs. The most widely used indicators of mapping are mapping of agents, mapping of core processes (activities), marketing channels, and flows of products. For the governance analysis, the researchers have generally used the list of constraints, quality standards, rules and regulations, and accession to markets, technology, and finance as a sort of indicator/outcome. The demand and supply conditions of the product have been investigated with the statistical data on area sown, yields, number of farmers, export and import statistics, and prices at the national and international levels. The end market analysis has been researched with the data of national and international market analysis. The last tool is SWOT analysis which is known as a general tool that is widely used for the VC analysis with the well-known SWOT indicators.

The second important area of interest in VC analysis is economic and financial analysis. Value-added analysis is an upper research area in the economical side of the VC analysis. Consumer behavior and financial analysis are also included in the case studies although they have some relatively minor importance. Value-added analysis consists generally of the share of marketing and profit margin by agents, cost of intermediate inputs, total output value, net value added, gross value added, and cost of fixed capital, respectively. The research on the consumer behavior of the VC has been focused on preferences of the consumers, attitude, behavior, and value of the consumers. Financial analysis is another area of interest in the VC analysis which is generally concentrated on cost of intermediate inputs, total output value, net present value, internal rate of return, cash flows, cost of fixed assets, and break-even point.

The social analysis of VC is an important aspect which covers people with social inclusivity on the basis of people's different gender and age groups. It also includes how much employment was created with decent working conditions.

Environmental analysis tries to evaluate the environmental effects of the VC which explains the effects on human health, resource depletion, and ecosystem quality, mainly resulting with the decline in biodiversity. Environmental analysis of VCs has been done by using different approaches such as hot spot analysis, environmental assessment, and life cycle assessment. Hot spot analysis uses qualitative data while the other assessments are subject to quantitative data analysis. Environmental analysis has some limited attention by the researchers, while hot spot analysis is being more preferred.

It is important to underline the tools and outcomes of the selected VCs. In other words, the agricultural VC means a set of goods and services needed for an agricultural product to move from the farm to the consumers. The chains contain various activities and actors involved, from the production process to delivery of a product to the market. There are different approaches that consider capturing the meaning of the agricultural VC (Rillo & Nugroho, 2016). Considering the issue in terms of added value in the agricultural food VC, the VC is a series of value-adding processes which flow across many companies and create products and services which are suitable to fulfill the needs of customers. Each step in the chain, from basic inputs to consumer goods, serves as a link or stage in the VC.

The guidelines provide various indicators or research questions for guiding the collection of data for VC analysis:

- ILO includes an increased employment and income as outcomes of VC development.
- UNIDO considers that implementation of the guidelines will result in VC with a greater possibility of obtaining positive impacts on poverty and gender equity.
- The guidelines by FAO emphasize the economic development aspects of VC development.

- CIP considers that creating trust and relationships between market chain actors is a necessity for being successful.
- M4P considers improving the environment in which the smallholders and other chain actors produce and market agricultural products as the basis for achieving VC development.
- VCA4D provides decision-makers with evidence-based information to feed sustainable development strategies.
- GFU presents stages and good practices for VC development of neglected and underutilized species.

Most of the studies on agricultural VCs are case studies, taking place in low- and middle-income countries. It is observed that, principally, qualitative data from surveys conducted for the studies are used, while quantitative evaluations are mostly based on cross-data in order to analyze the chains. Moreover, in some cases, chain analyses are only a description of the consecution of the steps. Even if this is useful for mapping actors and processes, it might not be adequate for guiding the actions of agencies and international institutions involved with policy making. Therefore, we need a demand-driven approach, considering the main agents of the agricultural VC such as consumers, supermarkets, retailers, input suppliers, and farmers for modelling agent interactions and behavior throughout the VC.

Every VC has its own unique factors and considerations. Four guidelines that were the most comprehensive in the design of VC were selected for this study. These guidelines were VCA4D, ACIAR, GTZ/GIS, and FAO (developing sustainable food VCs) that provide a comprehensive collection of tools and concepts related to VCs. These guidelines were complemented with others: M4P and GFU that provide a comprehensive collection of tools and concepts related to VC.

The study shows that when approaching VC selection from an economic, social, and environmental perspective, thereby facilitating a successful start to the VC development cycle, it can be understood that by combining production practices with the demand situation in agricultural markets, and constraints in the downstream of the VC, VCs are therefore vital for agricultural development.

Acknowledgments This study was prepared as part of the research Project-BIOVALUE: Fork-to-farm agent-based simulation tool augmenting BIOdiversity in the agri-food value chain-supported by the European Commission Grant Agreement No:101000499. The authors are grateful to the European Commission for granting support.

References

Anonymous. (2008). *M4P-making value chains work better for the poor: A toolbook for practitioners of value chain analysis, Version 3*. Making Markets Work Better for the Poor (M4P) Project, UK. Department for International Development (DFID). Agricultural Development International.

Bacon, C. (2005). Confronting the coffee crisis: Can fair trade, organic, and specialty coffees reduce small-scale farmer vulnerability in northern Nicaragua? *World Development, 33*(3), 497–511.

Bellu, L. G. (2013). *Value chain analysis for policy making, methodological guidelines and country cases for a quantitative approach*. FAO.

Bernet, T., Thiele, G., & Zschocke, T. (2006). *Participatory market chain approach (PMCA)-user guide*. International Potato Center.

BIOVALUE – D2.1. (2022). *Deliverable 2.1 – Reviewing of agro-food value chain tools*.

Clay, P. M., & Feeney, R. (2019). *International Food and Agribusiness Management Review, 22*(1), 31–46.

Collins, R. C., Dent, B., & Bonney, L. B. (2016). *A guide to value-chain analysis and development for overseas development assistance projects*. Australian Centre for International Agricultural Research.

Donovan, J., Franzel, S., Cunha, M., Gyau, A., & Dagmar, M. (2015). Guides for value chain development: A comparative review. *Journal of Agribusiness in Developing and Emerging Economies, 5*(1), 2–23. https://doi.org/10.1108/JADEE-07-013-0025

European Commission. (2018). *Value chain analysis for development (VCA4D), Methodological Brief. Frame and tools, key features of experts' work. Version 1.2*. https://europe.eu/capacity4dev/value-chain-analysis-for-development-vca4d-/documents/methodological-brief-v12.

FAO. (2014). *Developing sustainable food value chains – guiding principles*. FAO.

Haggblade, S. (2007). Subsector supply chains: Operational diagnostics for a complex rural economy. Chapter 15. In S. Haggblade, P. Hazell, & T. Reardon (Eds.), *Transforming the rural nonfarm economy* (pp. 352–378). Johns Hopkins University Press.

Henning, R., Donahue, N., & Brand, M. (2008). *End market research toolkit: Upgrading value chain competitiveness with informed choice*. USAID_AMAP-BDS.

ILO. (2021). Value chain development for decent work. A systems approach to creating more and better jobs. https://www.ilo.org/wcmsp5/groups/public/%2D%2D-ed_emp/%2D%2D-emp_ent/%2D%2D-ifp_seed/documents/publications/wcms_434362.pdf

Kaplinsky, R. (1999). Globalisation and unequalization: What can be learned from value chain analysis. *Journal of Development Studies, 37*(2), 117–146.

Lundy, M., Gottret, M. V., Ostertag, C., Best, R., & Ferris, S. (2007). *Participatory market chain analysis for smallholder producers*. International Centre for Tropical Agriculture.

Nang'ole, E. M., Mithöfer, D., & Franzel, S. (2011). *Review of guidelines and manuals for value chain analysis for agricultural and forest products*. ICRAF Occasional Paper No. 17. World Agroforestry Centre.

Rillo, A. D., & Nugroho, S. A. (2016). *Promoting agricultural value chain integration in Central Asia and the Caucasus*. Policy Brief, No: 4: 1–8.

Schneemann, J., & Vredeveld, T. (2016). *Guidelines for value chain selection: Integrating economic, environmental, social and institutional criteria*. Deutsche Gesellschaft für Internationale Zusammenarbeit (GIZ) GmbH.

Silva, C. A., & de Souza Filho, H. M. (2007). *Guidelines for rapid appraisals of agrifood chain performance in developing countries*. FAO.

Stoian, D., Donovan, J., Fisk, J., & Muldoon, M. (2012). Value chain development for rural poverty reduction: A reality check and a warning. *Enterprise Development & Microfinance, 23*(1), 54–69.

UNIDO. (2011). *Pro-poor value chain development: 25 guiding questions for designing and implementing agroindustry projects*. United Nations Industrial Development Organization (UNIDO).

Vermeulen, S., Woodhill, J., Proctor, F. J., & Delnoye, R. (2008). *Chain-wide learning for inclusive agrifood market development: A guide to multi-stakeholder processes for linking small-scale producers with modern markets*. International Institute for Environment and Development/Wageningen University and Research Centre.

Weinberger, K., & Lumpkin, T. A. (2007). Diversification into horticulture and poverty reduction: A research agenda. *World Development, 35*(8), 1464–1480.

Will, M. (2008). *Promoting value chains of neglected and underutilized species for pro-poor growth and biodiversity conservation, guidelines and good practices*. Global Facilitation Unit for Underutilized Species.

A Cluster Analysis Study Based on Financial Indicators, as a Competitiveness Strategy of the Brewery Industry in Greece

Maria Tsiouni, Christos Konstantinidis, Stamatis Aggelopoulos, and Spyridon Mamalis

1 Introduction

Approximately 1.91 billion hectoliters of beer were produced worldwide. The consumption of beer is widespread throughout the world, and in terms of beer production, China, the USA, and Brazil lead the world. Globally, beer is the most popular alcoholic beverage, and it ranks third overall in popularity behind water and tea. The European economy has benefited greatly from the beer industry. The total beer production in Europe for 2021 was 363.122 Hl (European Beer Trends, 2021). The brewery industry generates an estimated 2.3 million jobs. One brewery job creates an average of 17 additional jobs: 2 in supply and agriculture, 2 in retail, and 13 in bars, pubs, cafes, and restaurants (European Beer Trends, 2021).

The production of beer has been growing for several consecutive years, and this will not change. According to reports, many brands were developing new tastes, changing consumers' lifestyles, and maturing the markets, giving the industry an opportunity for growth (Swinnen, 2011).

In Greece, the brewing industry is one of the most important in terms of performance, employment, revenue, added value, and revenues. Concentration is high as a few large companies hold a large share of the market, which results in an intense competition that makes entry challenging for other smaller companies.

Beer consumption is influenced significantly by seasonality, which is one of the main factors influencing demand on the market for this product. Beer consumption

M. Tsiouni (✉) · S. Aggelopoulos
Department of Agriculture, International Hellenic University, Thessaloniki, Greece

C. Konstantinidis
School of Business Administration, International Hellenic University, Serres, Greece

S. Mamalis
Department of Management Science and Technology, International Hellenic University, Kavala, Greece

K. Mattas et al. (eds.), *Value Chain Dynamics in a Biodiverse Environment*, Cooperative Management, https://doi.org/10.1007/978-3-031-49845-9_7

increases during the summer months, largely because tourist traffic increases during this time, according to IOBE data (2022). Since tourism traffic was reduced as a result of the COVID-19 pandemic, as well as the closure of catering establishments, the sector saw significant declines in sales between 2019 and 2020 as a result of the seasonality of consumption and the COVID-19 pandemic.

According to IOBE (2022), the sector has shown signs of recovery over the course of the last year with particularly strong returns on key financial indicators. These indicators are a key tool for evaluating efficiency and competitiveness. The study of the financial indicators of a sector is therefore of particular importance both academically and economically, as it provides an important tool for determining the financial position of companies, determining the competitive conditions in the sector, and choosing and establishing the appropriate strategy for competitiveness and sustainability in the sector. The brewery sector is a key pillar of the Greek economy, showing significant growth even during economic recessions due to investment and business activity both within Greece and abroad. Many Greek companies in this sector have good prospects for sustainable development as it contributes to the domestic economy (Sofianopoulou, 2022).

Today, in Greece there are big brewery manufacturing industries that have expanded their production, adding a variety of new trademarks emphasizing their characteristics and origins, as well as a small number of breweries that are active in the field of brewing. However, in the last few years, several “microbreweries,” which operate mainly at a local level, have penetrated the production sector.

The chapter’s main objective is to identify and analyze the finances of companies active in the brewery sector. This can evaluate the grade they adapt to changes. Identifying the problems and analyzing their efficiency will be accomplished by analyzing the financial profiles of the companies. The application of a typological analysis could determine the level of brewery industry competition.

2 Literature Review

Throughout the years, many authors have studied operational performance, competitiveness, and financial performance (Kroes & Manikas, 2014). In the beer industries, exploratory studies have been conducted in the past, but they have had limited scopes or sample sizes. Many studies focused on beer consumption (Thome et al., 2016; Colen & Swinnen, 2010) or on consumer behavior (Capitello & Todirica, 2021; Hecht et al., 2020). A study by Tse et al. (2016) evaluated distribution as a component of risk and disruption, but the exploratory sampling was completed using an opinion scale, not a compounded financial or operational indicator. A study by Golicic et al. (2014) examined the institutions that make up the supply chain and its operations. The effect of brand equity on the competitive environment was measured with an inverse causal relation focused on the firm’s financials (Kim & Chao, 2018).

Competition in the brewing industry is characterized by different precedents. Financial performance is correlated with cash cycles, and the firm's liquidity. These relationships are explored in the literature (Zanotti et al., 2018). The field of accounting closely relates to financial statement analysis. Financial data are recorded, analyzed, presented, and interpreted by accounting activities associated with the production and exchange of goods and services within corporations and other institutions (Jumady et al., 2017). Accounting provides a method for determining whether a business makes a profit as a result of its transactions. An accounting system can provide information about a company's health and its results. Ratio analysis provides indicators and symptoms that are associated with the surrounding conditions. For drawing basic economic conclusions, indicator analysis is a common way of assessing production units and their strategic development (Tsiouni et al., 2022b). Several researchers have studied the factors influencing a firm's financial and economic conditions. In addition to economic analysis, the literature emphasizes the importance of financial analysis (Rossi, 2014; Tsiouni et al., 2022b). Indicators that measure the performance of a company can be used to assess the company's condition by comparing it to other companies of the same industry and/or by looking at their evolution over time. Generally, three ratios are used to measure a company's performance: the liquidity ratio, solvency ratio, and profitability ratio (Tsiouni et al., 2022a).

Elumah and Shobayo (2018) examined the financial performance of companies in the brewery industry using financial ratios. The studied firms were in Nigeria and the study period was 2011–2015. Based on the results, the brewery industry generates profit and returns it to shareholders efficiently through the use of its assets. Moreover, there is relatively low financial risk in the industry, and managers manage their stocks efficiently.

Zanotti et al. (2018), using brewing as an example, examined the underlying assumptions that connect the financial and operational performance of firms to the competitive environment. A total of 214 brewing companies from more than 12 European economies have been sampled. Study results indicate that firms' financial performance is significantly related to the competitive construct of their industries, but not necessarily to their operational performance. In addition, the operational structure of a firm does not necessarily affect its financial performance.

Abgata et al. (2021) examined Nigerian quoted breweries' performance with the use of financial ratios. According to the researchers' findings, Nigerian breweries are using an optimal debt-equity ratio relatively, as evidenced by the significant positive relationship between the debt-equity ratio and financial performance.

Susellawati et al. (2022) analyzed the relationships between profitability, liquidity, leverage, and activity ratios in manufacturing companies in the food and beverages industry sector that traded on the Indonesian stock exchange during 2016–2020. A significant relationship was found between the ratios.

Tsiouni et al. (2022a) examined the wineries' final position using financial indexes. Results showed that large firms were more competitive and profitable.

Other researchers have focused on a variety of factors that influence a firm's financial and economic conditions. A set of performance indicators, focusing on the evolution of the company, was analyzed by Giacosa et al. (2016). To improve growth, profitability, and debt repayment capacity, entrepreneurs must understand their environment and, at the same time, have the ability to adopt measures. After determining the current starting position, the enterprises can commit themselves to take virtuous positions economically, financially, and in terms of firm development.

Based on the above, not many studies have been conducted on the financial identity of various units of the brewery sector, and no studies have been conducted on the Greek brewery industry. It is considered necessary to study the financial situation and competitiveness of Greek brewery companies of all sizes and industries in light of globalization and the conditions under which they operate, as well as the difficulties Greek companies face due to financial problems (Pazarskis et al., 2018).

3 Materials and Methods

The sample contained 59 brewery industries. A few large industries exist in Greece, while many smaller and medium-sized companies produce large volumes, have a wide distribution network, and have high export sales. As well as having their own estates and advanced equipment, a significant number of small producers penetrate the market through a dynamic distribution system and high-quality products, and have their own estates and advanced equipment. Secondary data were used in order to analyze the financial position of the brewery enterprises. The data were collected from the balance sheets. A hierarchical cluster analysis of companies was conducted using SPSS software based primarily on secondary data analysis through qualitative indicators.

A sample of 59 companies was classified according to their turnover in the years 2015–2019 into three major categories. As a result, the following categorization was made:

1. Large companies: Turnover over 5 million euros
2. Medium companies: Turnover between 2 and 5 million euros
3. Small companies: Turnover between ½ and 2 million euros

For each business unit, the following ratios were selected to describe its financial profile: liquidity ratios, activity ratios, profitability ratios, and financial leverage ratios. Financial ratio results helped us to construct a cluster typology for each business unit in the brewery sector in order to identify similar units-clusters, highlighting any similarities or differences based on indicators of their economic status and financial profile. Small, medium, and large enterprises all have a unique ranking, regardless of their initial size (Ferraris & Grieco, 2015). Based on an analysis of financial indicators, business typologies were developed using hierarchical cluster analysis. Using Ward's criterion and Euclidean squares, clusters were formed (Hair et al., 1995; Sharma, 1996). In order to ensure that the results of the cluster analysis are stable according to the order in which companies were entered into the analysis, PermuCLUSTER 1.0 was used. The ratios that were used included the following:

acid test ratio (Y1), working capital to assets ratio (Y2), cash ratio (Y3), inventory turnover ratio (Y4), receivable turnover ratio (Y5), return on assets ratio (Y6), return on net worth ratio (Y7), and debt-to-assets ratio (Y8).

4 Results

The results of the analysis are presented in the following tables (Tables 1, 2, and 3):

Table 1 Financial ratio analysis during 2015–2019

	2015	2016	2017	2018	2019
Acid test ratio					
Large companies	5.33	9.08	12.04	10.13	7.69
Medium companies	2.04	5.97	6.87	9.81	4.97
Small companies	2.06	4.38	3.24	4.11	3.31
Working capital/turnover ratio					
Large companies	−0.15	0.25	0.47	0.98	0.59
Medium companies	0.01	0.20	0.13	0.19	0.38
Small companies	−0.08	0.19	0.14	0.18	0.22
Cash ratio					
Large companies	0.03	1.04	1.02	1.37	2.03
Medium companies	0.07	0.04	1.01	1.04	1.06
Small companies	0.11	0.10	0.92	1.10	1.08
Inventories turnover ratio					
Large companies	4.16	2.96	3.95	2.88	2.85
Medium companies	3.56	2.01	2.23	2.15	1.92
Small companies	1.61	0.85	1.59	1.31	1.28
Receivables turnover ratio					
Large companies	4.96	2.76	2.80	3.76	1.74
Medium companies	3.12	2.84	2.07	2.94	2.87
Small companies	3.35	2.73	2.98	2.94	2.92
Return on assets (ROA)					
Large companies	−0.01	−0.84	−0.51	0.02	0.06
Medium companies	−0.75	−0.82	−0.52	−0.03	0.03
Small companies	−0.37	−0.18	−0.01	−0.02	0.01
Return on net worth					
Large companies	−1.11	−1.03	0.00	0.19	0.21
Medium companies	−0.05	−0.07	0.23	0.02	0.08
Small companies	−0.03	−0.04	0.03	−0.04	0.04
Debt-to-assets ratio					
Large companies	0.65	0.68	0.98	1.61	1.68
Medium companies	0.51	0.56	0.54	0.57	1.56
Small companies	0.51	0.58	0.63	0.69	0.98

Table 2 Description of clusters

Clusters	Number of brewery companies	Percentage (%)
C_1	19	32.2
C_2	23	38.9
C_3	17	28.9

Table 3 Profile of clusters according to financial ratios

Clusters	Acid test ratio Y_1	Working capital to assets ratio Y_2	Cash ratio Y_3	Inventory turnover ratio Y_4	Receivable turnover ratio Y_5	Return on assets ratio Y_6	Return on net worth ratio Y_7	Debt-to-assets ratio Y_8
C_1	3.009[a]	0.180[a]	0.087[a]	3.28[a]	3.95[a]	−0.016[a]	0.467[a]	0.663[a]
C_2	5.325[b]	0.228[b]	0.042[b]	2.98[b]	2.86[b]	0.018[b]	−0.012[b]	0.651[a]
C_3	8.022[b]	0.194[a]	0.039[b]	1.97[c]	2.98[c]	0.011[b]	0.001[c]	0.548[a]

Based on the Dunnett T3 and Tukey tests, the columns with different letters exhibit significant differences by a significance level of * = 0.05

Based on Table 1, large companies have satisfactory prices, according to the acid test ratio (prices between 5.33 and 12.24) which applies to medium businesses as well (prices between 2.04 and 9.81). According to the working capital to total assets ratios, large companies maintained satisfactory liquidity reserves throughout the period 2015–2019. Generally, working capital ratios to total asset prices range from 0.11 to 0.38 for medium companies and −0.08 to 0.22 for small ones. Based on the cash ratio, it can be seen that all three categories of companies are showing a marginally downward trend, except for the year 2019, in which all categories of companies are showing an increase. For large companies, the inventory turnover ratio is higher (prices between 0.03 and 2.03) than medium (0.04–3.56) and small businesses (0.10–1.61), which may be related to the quite high profitability of large-sized enterprises due to small stockholdings. In light of the receivables turnover ratio values, medium-sized businesses (prices from 2.84 to 3.12) appear to have the best record of collecting receivables, followed by small (prices between 2.73 and 3.35) and large companies (prices between 1.74 and 4.96), suggesting that the different corporations have different lending policies. The return on assets ratio is negative for all companies for the years between 2015 and 2018. In 2019, it is positive for all categories of companies. As a result of this index value, the company's assets are inefficiently used. For large, medium, and small enterprises, the average of return on net worth ratio for the years 2015–2016 is negative, while for the years 2017–2019, the mean value of the price is positive. Low and negative prices indicate a survival problem for businesses. Negative index values indicate that equity has been used inefficiently. A higher debt-to-assets ratio is found for large enterprises (0.65–1.68) than for medium enterprises (0.51–1.56) or small enterprises (0.51–0.98). Large companies' higher index values indicate financial insecurity in the event of a failure, which poses a problem for business leaders. The risk for medium and small companies, on the other hand, is lower as prices are lower.

According to Table 2, the hierarchical analysis of the sample revealed three clusters of business units in the brewery industry, with 19 companies in the first (32.2%), 23 companies in the second (38.9%), and 17 companies in the third (28.9%) (Table 2).

According to Table 3, the acid test ratio shows differences between the first class (3.009) and the second and third (5.325 and 8.022, respectively). For the working capital ratio to total assets, the first (0.180) and third (0.194) cluster companies differ from the second (0.228) cluster companies. The cash ratio differentiates the companies in the second (0.042) and third (0.039) clusters from the companies in the first (0.087) cluster, while for the inventory turnover ratio, there are differences in all clusters of the sample. For the receivables turnover ratio, the results are also similar, since there are differences between companies within all clusters. There is also a difference between the first cluster (−0.016) and the second and third (0.018 and 0.011, respectively) in terms of the return on assets ratio. According to the return in net worth ratio, there is differentiation among all clusters. And finally, according to the debt-to-assets ratio, there is no differentiation among clusters.

5 Discussion and Conclusions

Based on the ratio analysis, we can conclude that the industry can invest in growth and profitability because, in some cases, the return to shareholders is worthwhile. It is also evident from analyzing the indicators that the industry is growing, which means it is likely to attract investors in the future. According to our liquidity ratios, the sector has satisfactory liquidity. The capital structure and solvency indicators conclude that in large companies, stocks circulate rapidly, demonstrating a high level of solvency.

Businesses with small or medium-sized revenues may have a harder time meeting their current obligations and coping with potential losses. It is generally perceived that the indicators measuring the efficiency of asset utilization in large and medium-sized enterprises are stable, but an excessive overinvestment in capital relative to sales could lead to failure to meet obligations in the future. A high debt-to-assets ratio affects large companies more than medium and small ones, which must be addressed by increasing the firms' liquidity to avoid debt service difficulties.

Beverage production is a very important source of income for our country, whether it comes from larger breweries or from smaller ones. Brewery industry businesses operate in an unfavorable business environment characterized by a prolonged economic recession, increased tax burdens, and a shrinking market.

The increase in annual sales and the creation of a strong Greek brand will lead to the development of the sector. Moreover, the reduction of the tax will give rise to the opportunity costs of the brewery industry.

Increased repayment periods of loans should ensure the long-term survival of the sector's companies. Small and medium-sized businesses need access to low-interest loans through strengthening and recapitalizing credit institutions. In this way, small businesses will be able to respond to new opportunities, change with changing markets, and adjust rapidly to economic upswings, in contrast with large businesses.

References

Agbata, E., Osingor, S., & Ezeala, G. (2021). Effect of financial ratios on firm performance: Study of selected brewery firms in Nigeria. *International Journal of Trend in Scientific Research and Development, 5*(5), 1664–1671. www.ijtsrd.com/papers/ijtsrd45177.pdf

Capitello, R., & Todirica, I. C. (2021). Understanding the behavior of beer consumers. *Case Studies in the Beer Sector*, 15–36. https://doi.org/10.1016/b978-0-12-817734-1.00002-1

Colen, L., & Swinnen, J. F. (2010). Beer drinking nations – the determinants of Global Beer Consumption. *SSRN Electronic Journal*. https://doi.org/10.2139/ssrn.1752829

Elumah, L. O., & Shobayo, P. (2018). Performance analysis of Nigerian brewery industry. *Binus Business Review, 9*(1), 47. https://doi.org/10.21512/bbr.v9i1.4047

European Beer Trends. (2021). *European beer trends statistics report* | 2022 Edition.

Ferraris, A., & Grieco, C. (2015). The role of the innovation catalyst in social innovation—An Italian case study' sinergie. *Italian Journal of Management, 33*, 127–144.

Giacosa, E., Culasso, F., Mazzoleni, A., & Matteo, R. (2016). A model for the evaluation trends performance in small and medium enterprise. *Corporate Ownership and Control, 13*(4), 389–402.

Golicic, S. L., Davis, D. F., Davis-Sramek, B., & McCarthy-Byrne, T. M. (2014). *Institutional pressures and relationships in the wine supply chain* (8th ed., pp. 1–16). Academy of Wine and Business Research.

Hair, J., Anderson, R., Tatham, R., & Black, W. (1995). *Multivariate data analysis with readings* (4th ed.). Prentice-Hall International.

Hecht, A. A., Perez, C. L., Polascek, M., Thorndike, A. N., Franckle, R. L., & Moran, A. J. (2020). Influence of food and beverage companies on retailer marketing strategies and consumer behavior. *International Journal of Environmental Research and Public Health, 17*(20), 7381. https://doi.org/10.3390/ijerph17207381

IOBE. (2022). http://iobe.gr/

Jumady, E., Pattiware, A., Utami, M., & Ishak, R. (2017). Hubungan Antara Profitabilitas Dan Leverage Dengan Stock Return Pada Industri Manufaktur Dibursa Efek Indonesia. *Jurnal Ilmiah Bon-gaya, 2*(2), 25–32.

Kim, R., & Chao, Y. (2018). The effect of country of origin on consumer-based Brand Equity (CBBE) of Colombian consumers: An empirical investigation of Samsung vs. Huawei Brands. *Journal of International Studies, 11*(3), 70–81. https://doi.org/10.14254/2071-8330.2018/11-3/6

Kroes, J. R., & Manikas, A. S. (2014). Cash flow management and manufacturing firm Financial Performance: A longitudinal perspective. *International Journal of Production Economics, 148*, 37–50. https://doi.org/10.1016/j.ijpe.2013.11.008

Pazarskis, M., Alexandrakis, A., Vogiatzoglou, M., & Drogalas, G. (2018). Measuring the effect of mergers in Greece with the use of financial ratios: A bootstrapped approach. *Theoretical Economics Letters, 8*, 421–431. https://doi.org/10.4236/tel.2018.83030

Rossi, M. (2014). The use of capital budgeting techniques: An outlook from Italy. *International Journal of Management Practice, 7*(4), 297–312.

Sharma, S. (1996). *Applied multivariate techniques*. Wiley.

Sofianopoulou, N. (2022). *Economic analysis of brewing in Greece during the latest financial crisis*. Master Thesis, Agricultural University of Athens, MBA FOOD & AGRIBUSINESS, Athens.

Susellawati, D. P., Sari, S. D. N., & Fadlilah, A. H. (2022). The effect of profitability, liquidity, leverage, and activity ratios on dividend policy in manufacturing companies in the food and beverage industry sector listed on the Indonesia Stock Exchange in the 2016-2020 period. *Journal Mantik, 6*(2), 1365–1375. https://doi.org/10.35335/mantik.v6i2.2494

Swinnen, J. F. (2011). *The economics of beer*. Oxford University Press. https://doi.org/10.1093/acprof:oso/9780199693801.001.0001

Thome, K. M., Pinho, G. D., Fonseca, D. P., & Soares, A. B. (2016). Consumers' luxury value perception in the Brazilian premium beer market. *International Journal of Wine Business Research, 28*(4), 369–386. https://doi.org/10.1108/IJWBR-09-2015-0043

Tse, Y. K., Matthews, R. L., Tan, K. H., Sato, Y., & Pongpanich, C. (2016). Unlocking supply chain disruption risk within the Thai beverage industry. *Industrial Management & Data Systems, 116*(1), 21–42. https://doi.org/10.1108/IMDS-03-2015-0108

Tsiouni, M., Konstantinidis, C., Pavloudi, A., & Giovanis, N. (2022a). Financial ratio and efficiency analysis as a competitive advantage of wine manufacturing firms. the case of Greece. *Theoretical Economics Letters, 12*(01), 229–239. https://doi.org/10.4236/tel.2022.121013

Tsiouni, M., Gourdouvelis, D., Aggelopoulos, S., & Siggia, D. (2022b). Improve the financial management practices in goat farms with the study of financial ratios. The case of Greece. *International Journal of Managerial and Financial Accounting, 14*(2), 184. https://doi.org/10.1504/ijmfa.2022.122225

Zanotti, C., Reyes, F., & Fernandez, B. (2018). Relationship between competitiveness and operational and financial performance of firms: An exploratory study on the European Brewing Industry. *Intangible Capital, 14*(1), 99. https://doi.org/10.3926/ic.1104

Climate Change and the Role of Governance in the Value Chain Sustainability of Carob Flour in Rethymno, Crete, Greece

Andreas Vavvos, Iosif Kafkalas, Charalambos-Nikolaos Piteris, and Kondylia Skrapaliori

1 Introduction

The carob tree, officially named *Ceratonia siliqua L.,* is a plant that has grown and been planted in most countries of the Mediterranean basin since antiquity, adapted in mild and dry places with poor soils (Batlle, 1997). For Greece and especially the presented area of Crete in this chapter, it has been an important plant for economic, cultural, and social sustainability. Locals mention its characterization as "black gold" in tradition, which is also mentioned in nearby Mediterranean areas like Cyprus (Papaefstathiou et al., 2018), while especially widespread in Crete is the characterization of carob as the "chocolate of the poor" and its nutritional importance in times of hardship, like during the Second World War occupation and the "Great Famine" (1941–1944).

The area of interest is a semi-mountainous region, lying between the two main mountain masses of Crete: Mount Ida or as the locals call it "Psiloritis" (2456 m) and the White Mountains or "Madares" (2453 m). In an island mostly covered by olive trees, carob trees in Crete have traditionally been interplanted in olive groves with low-intensity farming systems or spread naturally in terrain that is more difficult to cultivate. The value chain (VC) of carob pods and carob flour has historically been important for the area, either for human consumption in low quantities or as a significant part of animal feed.

A. Vavvos (✉)
Department of Psychology, University of Crete, Crete, Greece

I. Kafkalas
Department of Economics, University of Crete, Crete, Greece

C-N. Piteris · K. Skrapaliori
Region of Crete, Crete, Greece

K. Mattas et al. (eds.), *Value Chain Dynamics in a Biodiverse Environment*, Cooperative Management, https://doi.org/10.1007/978-3-031-49845-9_8

Carob pod production declined precipitously because of harvesting abandonment, the destruction of carob trees for firewood, and land-use changes (EU subsidies for olive trees and their oil, which led to the "olive monoculture"). The focus on more intensive olive oil production together with its drop in price has reduced the importance of carob tree cultivation and led to the disappearance of many crucial links of its value chain, previously supported by traditional community practices. The rising price of carob products in the last decade, followed by a rise in the acknowledgement of their nutritional and environmental value, has revived interest in carob tree cultivation by local communities, which now look back to forgotten traditions together with innovative practices for enhanced productivity. In the past 5 years, the carob flour VC's development has been valorized by scientific capital that has focused on carob flour's unique nutrients and its fit with the Cretan and Mediterranean diet (Barros & Delgado, 2022), rendering it essentially a "superfood". Carob consumption has also ascertained benefits in fighting obesity and related metabolic disorders (Gioxari et al., 2022).

Carob trees' cultivation has also been highlighted in the literature for its importance in addressing climate change challenges and contributing as a vital component of environmental sustainability. Soil degradation is an issue of global concern, and in Europe, one of the leading causes of soil degradation is unsustainable agricultural practices (Sollen-Norrlin et al., 2020). Carob trees can survive in very precarious conditions, with its rich root system retaining and protecting the soil from erosion (Tzatzani & Ouzounidou, 2023). Carob trees are also well adapted to marginal soils, both calcareous and rocky, and low rainfall (Correia & Pestana, 2018), which makes them capable of surviving droughts and on soils that are inadequate for other types of cultivation.

In this study, using a variety of archival sources, regional government statistics, interviews, and workshop results, we are trying to explore how the value chain of carob and carob flour can be a promising alternative to the monoculture of olive oils and the water-intensive avocado cultivation in Crete. Participants were mainly producers, farmers, businessmen, and scientists that are involved in the carob flour value chain.

2 The Value Chain of Carob Flour

In order to depict the local value chain of carob, we used a plethora of alternative ways to interact with local stakeholders. The first step was the conduct of 17 interviews with important local stakeholders in order to identify the reference variable; the drivers of change and the relevant components of each driver will be extracted and synthesized from the interviews by the region team. The main question was how the reference variable is influenced by each driver of change. With snowball sampling, we recruited participants with extensive knowledge of the value chain and the land-use system that belonged to one of the following four categories: farmers/producers, extension officers, managers, and researchers.

Carob trees produce carob pods, which in turn can be used in the production of a wide variety of products, from edible products to cosmetics, and even in some types of film products (Pérez et al., 2021). However, in our value-chain analysis we focus on the carob flour as the main product purposely produced by the farmers and the most important by-products that local farmers in the area of analysis usually expect to profit from. Visualization of the value chain is depicted in the following figure (Fig. 1).

Almost 80–90% of the carob production pertains to the pulp, and only 10–20% refers to the seeds. The higher-value carob seeds are extracted, exported, and processed in Italy and other countries and are ultimately utilized in gastronomy. The carob legumes are processed for both human and animal consumption in modern mills with patented techniques for processing the ripe, dried, or toasted legumes into flour, chips, powder, coffee, tea, beauty products, etc. The value chains of carob flour and animal feed are very competitive value chains, but the most profitable use of carob is in the pharmaceutical industry.

3 Carob Production Against Climate Change

Twenty years from today, the literature predicts that the Mediterranean region and Crete will suffer from decreasing water reserves, more frequent meteorological and hydrological droughts, and less frost (Georgoulias et al., 2022; Tapoglou et al., 2019). These conditions raise the advantages of carob trees as resilient, adaptive, viable in poor soils (rocky) and requiring almost no care to thrive. Carob trees have low requirements concerning orchard management and are suitable for the part-time farming practices of small holders. Sheep and goat herding along with carob

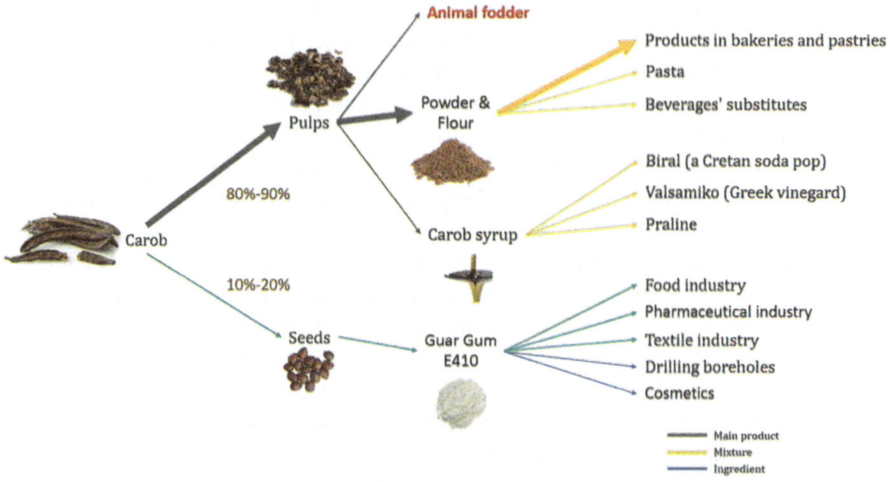

Fig. 1 The value chain of carob flour in the area under research

tree cultivation are complementary activities that improve the environment by remediating pollution, preventing fires, and naturally enhancing soil fertility (Papanastasis et al., 2009).

Disadvantages include the fact that pod yields are uneven and economic yields are tied to the price of carob seeds which are exported and processed abroad. Inconsistency and decline in yields are also tied to demographic changes in the mountainous and semi-mountainous villages, leading to loss of knowledge from traditional practices, and finally, international competition from countries where carob pod production is an intensive activity. Other studies also propose carob as an excellent alternative to intensive crops and against climate change. According to Correia and Pestana (2018), carob trees are very potent in absorbing large quantities of CO_2, while according to Issaoui et al. (2021), another sustainable characteristic of carob production is that the food processing of carob pods does not generate any waste.

4 Vulnerability Analysis and Drivers of Change

During the interviews and workshops, participants were asked about the weakest links of the value chain and the potential to strengthen their processes against future risks. The demographic changes, the land-use alternatives, and the rising temperatures have been indicated as the drivers with a sharp increase in the last 20 years.

According to their responses, the most dominant driver of change for the carob production in the area of analysis is demographics, especially the flow of the local population to nearby cities, reducing available workforce and the necessary part of the local population that is needed to stay and train in the practices of carob production and processes. Traditional processes were not only specific skills to learn and perform, but more of a holistic cultural heritage that includes long-term relationships with the community and interaction with other people and practices in the area. Participants believe that demographic changes had probably the most significant effect in the value chain in the past, and they predict that in the future, this effect will grow even stronger.

The second significant factor restraining carob production in the area is the competing alternative cultivations, raising the opportunity cost of land use, mainly in favour of olive-tree cultivation. As long as local producers focus on monocultures of olive trees, they focus their motivation on the increase of productivity and efficiency in this area, reducing interest from the traditional polyculture practices that included carob production and its association with other sectors of production in the area. For example, the lack of interest in carob production has also reduced animal feed available to local shepherds, weakening the interdependence of local value-chain participants and increasing the dependence on imported feed that played a role in increasing the cost of production and risk vulnerability.

Finally, wildfires are considered non-existent as a risk factor in the area. Overexploitation and soil physical degradation have remained more or less constant in the last few decades, whereas pollution has exhibited a slight increase, which, according to the stakeholders, has not affected the carob production. The climate change drivers are described in the next section, although most of the participants do not seem to have a clear perception of their importance.

5 Drivers of Environmental Change

In this step, we present the different environmental values of the value chain in each stage (production, processing, marketing/distribution, and consumption) as they were indicated by the results of interviews and workshops. In the production stage, the value chain contributes to the high fixation of CO_2, climate change mitigation, and mitigation of soil erosion, while it is beneficial for the biodiversity in the area. There is a documented need for policy action regarding crop production in agro-silvo-pastoral systems. In the processing stage, the value chain contributes to agro-ecology and agricultural biodiversity and to the reinforcement of traditional sustainable farming practices. In the third stage of marketing and distribution, it is known that food transport accounts for <4% of food-related greenhouse emissions, while inflation in fuel prices is troubling and economically difficult to sustain. Last but not least, regarding consumption, it is well known that food consumption is a driver for environmental pressure.

Particularly, the main drivers are the following:

1. *Precipitation*: Carob pod/bean production depends on favourable natural rainfall cycles in early autumn and spring. Carob trees grow in the arid and drought-prone soils, but pod production can be negatively affected by very low rainfall in September and the late spring months. The driver was included due to the evolving climate change. To keep it simple, the adopted description was "PRECIPITATION concerns significant changes in rainfall from autumn to mid-spring (FAVOURABLE RAINFALL). Changes in this period have negative effects on the hydrological level (streams and aquifers), soil and vegetation. Sufficient rainfall in this period is important for carob germination, flower set, and pod production. Therefore, changes in this rainfall period can negatively affect carob trees, as well as other crops. Specifically, in this study we will focus on the prolonged reduction of precipitation during springtime."
2. *Temperature:* Trees and pod growth endure a temperature range from −5 to 50 °C. Adult trees suffer damages when temperatures fall below −4 °C and there is frost, and young trees are even more sensitive. In contrast, high temperatures up to 50 °C are well tolerated by carob trees. In the island of Crete, there is a low risk of vulnerability to low temperatures and low susceptibility and sensitivity of carob trees to low temperatures, but sensitivity to temperatures below −5 °C. The driver was included due to the evolving climate change. The adopted description was "Understanding TEMPERATURE as the increase of temperatures on average over the year. This results in milder winters and

warmer summers, which positively affect the vegetation, flowering and fruiting of the carob trees. An increase in mean temperature over the year is likely to have a positive effect on the production of carob pods."

3. *Extreme events:* Extreme southern winds break down trees and dislodge the carob pods/beans, impeding production. The area is prone to such weather conditions, and they have become more prominent. Frost and temperatures below 5 °C damage pods and trees, but this are a very rare occurrence in the MRL. Additionally, in the last 10 years the area experienced rapid rainfalls in winter. The driver was included due to the evolving climate change. The adopted description was "EXTREME EVENTS means changes in the intensity, frequency or timing of extreme weather events. For the carob-growing area in Rethymno, we focus on rapid rainfall in winter and heat waves in combination with strong southerly winds and prolonged hot conditions starting in spring that negatively affect carob trees."
4. *Wildfires:* The agro-pastoral land management system is associated with the prevention of wildfires. Wildfires are not prevalent on Crete, and carob trees burn less slowly than all other trees. The driver was included in case a future threat would be indicated by the participants, but it did not happen.
5. *Land-use and land-cover change:* Land-use changes are driven by profit margins (leading to a shift to the monoculture of olives), EU subsidies for olive farming, road construction and land fragmentation, and carob harvesting abandonment/uprooting and timbering to firewood. The adopted description was "Understanding LAND-USE and LAND-COVER CHANGE as abandonment of carob crops, uprooting of carob trees for the production of firewood and land-use change mainly for olive cultivation."
6. *Natural soil degradation:* Carob grows under different edaphic conditions. The trees grow and produce pods in marginal and calcareous soils. Soil properties in the area are sufficient for pod production in the area. The agro-pastoral land management system impacts positively via weed reduction and natural fertilization. The driver was included in case the perception of a group of participants would be different. The adopted description was "Understanding SOIL PHYSICAL DEGRADATION as loss of organic matter due to erosion, poor tillage, or excessive compaction. In the case of carob trees, there is no natural degradation, as carob trees have marginal nutrient requirements and pasture land use coexists. Carob cultivation is suitable for soils that have been degraded by other crops and could be used to upgrade them."
7. *Overexploitation of resources:* Carob pod production does not exploit the natural resources since it is a dry and organic farming crop. Carob trees have a positive impact on resources since their long rooting systems aid in preventing soil erosion. The driver was included in case the perception of a group of participants would be different. The adopted description was "Understanding SOIL PHYSICAL DEGRADATION as loss of organic matter due to erosion, poor tillage, or excessive compaction. In the case of carob trees, there is no natural degradation, but other neighbouring crops and land-use systems may affect the soil."

8. *Pests, diseases, and invasive species:* Pest bugs can damage pod production. Rodents destroy tree bark causing damage to trees and pod production. Production may be impacted much more if pests and more invasive species infiltrate. The danger of invasive species was mainly underlined by researchers and publications related to problems in carob productions in Sicily and south Italy. That was the main reason that the particular driver was included. The adopted description was "Understanding PESTS, DISEASES and INVASIVE SPECIES as spread of rodents, new diseases and species that have invaded Crete and affect carob trees and/or their pods."
9. *Pollution:* It is not clear if pollution impacts production of carob pods. Airborne pesticides may impact production, especially in certain areas where aerial spraying for olive trees was carried out in the past. The driver was not self-evident, and it was included in order to check the extension and the possible impact of such types of "indirect" pollution. The adopted description was "Understanding POLLUTION as the systematic degradation of natural resources due to intentional or unintentional human intervention. In the carob-growing areas of Rethymno, the pollution can be caused by aerial spraying, use of pesticides and/or elimination of processing products and production of other products (oil production or dairy production). Pollution can adversely affect carob pod production."
10. *Demographic changes:* Land/harvesting abandonment is due to ageing and declining population in mountainous and semi-mountainous areas, away from the urban centres (>10 km), and the farming occupation abandonment by the younger generation. Loss of knowledge regarding grafting and pruning is due to demographic changes; decades of harvesting abandonment has impacted pod production. The driver was self-evident for participants. The adopted description was "Understanding DEMOGRAPHIC CHANGES as population ageing and decline, and decline of agricultural activities in the mountainous areas of Rethymno. In addition, demographic changes have caused a reduction in the number of carob cultivators and a loss of significant knowledge for the development, collection and production of carob."

6 Carob Flour Value Chain and Governance

New governance structures are required according to participants, with regard to the need for carob tree cloning/breeding programmes, the reduction of the vegetative period (i.e. increase precocity in bearing), and the increase of pod and seed yield. New governance systems planning that will incentivize young farmers to engage in carob production are needed. The maintenance of agroforestry systems and the creation of new ones are considered important for maintaining the good condition of the soils, biodiversity, and the landscape, as well as ensuring financial support for the rural population that practices such activities. Strengthening the measure of integration of third-country nationals into the area is an indirect way of addressing the "demographic changes", which is the most important driver of change for the value chain.

The lack of know-how was repeatedly reported as a major problem for producers and local businesses. Support is needed for the adaptation and modernization of education, training, and employment systems to assist the area's youth. The Strategic Planning of the Region of Crete 2020–2023 includes the Strengthening of Lifelong Learning with Targeted Lifelong Learning Programs for all age groups in cutting-edge topics (indicatively, climate change and the circular economy), while the inclusion of corresponding topics in formal and non-formal education is also necessary.

The performance of the VC would be more effective, if policies that empower agricultural directorates, agronomists, and farming cooperatives were developed and implemented. Greater autonomy to Regional Policies should be given, so that the Region can design, implement, and finance in order to support the value chain in specific actions. A reforestation effort was implemented in order to increase the wealth of the villages in the area and develop them on site. However, farmers who took part in the afforestation of agricultural land (Regulation (EC) No 2080/92) programme that was implemented in 1996 in the area were not satisfied by the results.

The agribusinesses (flour mills, bakeries, and confectionery businesses) complained that funding or loans for new equipment are extremely difficult to attain. We realized that there is an astonishing lack of data regarding carob production, number of producers, and other relevant information. Institutions that collect and record information are urgently needed, in order to bolster guidance, support, and availability of directorates and extension officers, including record-keeping and research initiatives that can assist farmers and agribusinesses. Native carobs grow in predominantly forest and grassland ecosystems of Crete, with proven abilities to adapt to climate fluctuations and unchanged quality characteristics. However, the composition of the vegetation is related to the ownership regime, and this is not helpful in the value chain, as it results in the deforestation or removal of forest trees, as citizens attempt to avoid the claiming of these lands by the state.

Acknowledgements This study has received funding from the European Union's Horizon 2020 research and innovation programme under grant agreement No. 862739 (MOVING). This publication reflects only the authors' views. The European Commission is not responsible for any use that may be made of the information it contains.

References

Barros, V. C., & Delgado, A. M. (2022). Mediterranean diet, a sustainable cultural asset. *Encyclopedia, 2*(2), 761–777.

Batlle, I. (1997). *Carob tree: Ceratonia siliqua L. – Promoting the conservation and use of underutilized and neglected crops* (Vol. 17). Bioversity International.

Correia, P. J., & Pestana, M. (2018). Exploratory analysis of the productivity of carob tree (Ceratonia siliqua) orchards conducted under dry-farming conditions. *Sustainability, 10*(7), 2250.

Gioxari, A., Amerikanou, C., Nestoridi, I., Gourgari, E., Pratsinis, H., Kalogeropoulos, N., Andrikopoulos, N. K., & Kaliora, A. C. (2022). Carob: A sustainable opportunity for metabolic health. *Foods, 11*, 2154. https://doi.org/10.3390/foods11142154

Georgoulias, A. K., Akritidis, D., Kalisoras, A., Kapsomenakis, J., Melas, D., Zerefos, C., Issaoui, P. Z., Flamini, G., & Delgado, A. (2022). Climate change projections for Greece in the 21st century from high-resolution EURO-CORDEX RCM simulations. *Atmospheric Research, 271*, 106049.

Issaoui, M., Flamini, G., & Delgado, A. (2021). Sustainability opportunities for Mediterranean food products through new formulations based on carob flour (Ceratonia siliqua L.). *Sustainability, 13*, 8026.

Papaefstathiou, E., Agapiou, A., Giannopoulos, S., & Kokkinofta, R. (2018). Nutritional characterization of carobs and traditional carob products. *Food Science & Nutrition, 6*(8), 2151–2161.

Papanastasis, V. P., Mantzanas, K., Dini-Papanastasi, O., & Ispikoudis, I. (2009). Traditional agroforestry systems and their evolution in Greece. *Agroforestry in Europe: Current Status and Future Prospects*, 89–109.

Pérez, M. J., Moreno, M. A., Martínez-Abad, A., Cattaneo, F., Zampini, C., Isla, M. I., López-Rubio, A., & Fabra, M. J. (2021). Interest of black carob extract for the development of active biopolymer films for cheese preservation. *Food Hydrocolloids, 113*, 106436.

Silaev, A. A. (2017). Carob (Ceratonia siliqua): Health, medicine and chemistry. *European Chemical Bulletin, 6*(10), 456–469.

Sollen-Norrlin, M., Ghaley, B. B., & Rintoul, N. L. J. (2020). Agroforestry benefits and challenges for adoption in Europe and beyond. *Sustainability, 12*(17), 7001.

Tapoglou, E., Vozinakis, A. E., & Tsanis, I. (2019). Climate change impact on the hydrometeorological extremes in the island of Crete. *Water, 11*(3), 587.

Tzatzani, T. T., & Ouzounidou, G. (2023). Carob as an agrifood chain product of cultural, agricultural and economic importance in the Mediterranean region. *Journal of Innovation Economics Management*, I140–I121.

SPRINGER NATURE

GPSR Compliance

The European Union's (EU) General Product Safety Regulation (GPSR) is a set of rules that requires consumer products to be safe and our obligations to ensure this.

If you have any concerns about our products, you can contact us on ProductSafety@springernature.com

In case Publisher is established outside the EU, the EU authorized representative is:

Springer Nature Customer Service Center GmbH
Europaplatz 3
69115 Heidelberg, Germany

The manufacturer's authorised representative in the EU is Springer Nature Customer Service Centre GmbH, Europaplatz 3, 69115 Heidelberg, Germany. If you have any concerns regarding our products, please contact ProductSafety@springernature.com

Printed and bound by CPI Group (UK) Ltd, Croydon, CR0 4YY
01/10/2026
02225916-0012